Revision of *Ufens* Girault, 1911 (Hymenoptera: Trichogrammatidae)

Albert K. Owen

Revision of *Ufens* Girault, 1911 (Hymenoptera: Trichogrammatidae)

Albert K. Owen

UNIVERSITY OF CALIFORNIA PRESS
Berkeley • Los Angeles • London

University of California Press, one of the most distinguished university presses in the United States, enriches lives around the world by advancing scholarship in the humanities, social sciences, and natural sciences. Its activities are supported by the UC Press Foundation and by philanthropic contributions from individuals and institutions. For more information, visit www.ucpress.edu.

University of California Publications in Entomology, Volume 131
Editorial Board: Rosemary Gillespie, Penny Gullan, Bradford A. Hawkins, John Heraty, Lynn S. Kimsey, Serguei V. Triapitsyn, Philip S. Ward, Kipling Will

University of California Press
Berkeley and Los Angeles, California

University of California Press, Ltd.
London, England

Library of Congress Control Number: 2011937757
ISBN 978-0-520-09887-9 (pbk. : alk. paper)

CONTENTS

Acknowledgements

I would especially like to thank my major advisor, John D. Pinto. His insight, breadth of knowledge, professionalism and work ethic provided a constant source of inspiration. I would also like to thank my co-advisor, John Heraty, and laboratory research associate, Gary Platner. This work would not have been possible without their help, insight, and encouragment. My wife, Tania, showed good-natured tolerance at my long hours and provided an endless supply of support.

This work and my sanity also benefited from interactions with other colleagues at the University of California, Riverside, including: Matt Buffington, Roger Burks, Bryan Carey, Andy Carmichael, David Hawks, Michael Gates, Jeremiah George, Jung-Wook Kim, Johan Liljeblad, Jason Mottern, James Munro, Kris Tollerup, Serguei Triapitsyn and Doug Yanega. Many collaborators the world over helped me with insight and specimens, including: Gennaro Viggiani (Dipartimento di Entomologia e Zoologia Agraria dell'Universitá, Portici, Italy); Chris Burwell (Queensland Museum); Michael Gates, Eric Grissell, Terry Erwin and Michael Schauff (National Museum of Natural History); Gary Gibson and John Huber (Canadian National Insect Collection); Alejandro Gonzalez (Universidad Autónoma de Nuevo León); John LaSalle (Australian National Insect Collection); Naiquan Lin (Fujian Agricultural College, Fujian, China); John Noyes (The Natural History Museum, London); Lars Vilhelmsen (University of Copenhagen); Jim Woolley (Texas A & M University); and Robert Zuparko (California Academy of Sciences). I give a special thanks to Molly Hunter (The University of Arizona) for introducing me to the world of parasitic hymenoptera. This study was supported in large part by PEET grants from the National Science Foundation (DEB-9978150 and DEB-0730616; J. Heraty, PI). Phil Ward of the University of California, Davis and two anonymous reviewers improved the quality of this monograph with their comments.

Abstract

The first worldwide revision of the wasp genus *Ufens* Girault, 1911 (Hymenoptera: Trichogrammatidae: Oligositinae) is presented. *Ufens* is known to parasitize primarily hemipteran eggs and is a cosmopolitan genus most common in temperate and semi-arid regions such as the southwestern United States and Australia. Forty-three species are recognized here. Included in the revision are five species formerly in the genus *Ufensia*, herein synonymized, and 32 new species. In addition, seven species are removed from *Ufens* and placed in renewed combination in *Mirufens*. Because thirteen nominal species remain unidentifiable they are treated as *nomina dubia*. A worldwide key to species is provided. A preliminary phylogenetic hypothesis is presented utilizing both molecular and morphological data in maximum parsimony analysis. Molecular data, however, are limited to twelve of the recognized species. Thirty-seven morphological characters were utilized, both alone and together with molecular data. Due to greater confidence in molecular results and overall lack of resolution, the results of the paired-down molecular plus morphological analysis were utilized as a backbone for analyzing the complete morphological data set. Overall, results are inconclusive, with few relationships consistently recovered. There does appear to be a Holarctic clade, but even this result is tenuous.

Introduction

Scope of work. The genus *Ufens* Girault (Hymenoptera: Trichogrammatidae: Oligositinae) (Fig. 5) was erected in 1911 by A. A. Girault to account for the distinctness of a species described by Ashmead (1888) as *Trichogramma nigrum* (Girault 1911a). Its hosts are primarily hemipteran eggs (Pinto 1997). Twenty seven species of *Ufens* have been previously recognized (Lin 1994, Noyes 2002). Forty-three species are recognized here (Table 2). Included are five species formerly in the genus *Ufensia*, herein synonymized, and 32 new species. In addition, seven species are removed from *Ufens*, and placed in renewed combination in *Mirufens*. Because thirteen nominal species remain unidentifiable they are treated below as *nomina dubia*. The first phylogenetic hypotheses for *Ufens* are herein proposed, utilizing both morphological and molecular data. Molecular data, however, are limited to only twelve of the recognized species. Results are inconclusive, with few relationships consistently recovered. The only keys to species previously available were for the Palearctic (Nikol'skaya 1952, Nikol'skaya and Trjapitzyn 1987, Viggiani 1988) and India (Yousuf and Shafee 1987); a checklist of species was presented by Lin (1994). The worldwide key to *Ufens* species presented herein incorporates known geographical distribution, with the caveat that geographical distributions remain poorly known for most species. Although a cosmopolitan genus, *Ufens* is most common in temperate and semi-arid regions such as the southwestern United States and Australia (Table 3). The only major geographic area where *Ufens* appears to be poorly represented is South America, considering that only one known specimen has ever been collected there (*U. taniae*). Instead, in South America, *Zagella* appears to be the dominant trichogrammatid reared from hosts that commonly harbor *Ufens* in other regions (Triapitsyn 2003).

Trichogrammatidae. *Ufens* is a member of the Trichogrammatidae, whose members are solitary or gregarious idiobiont endoparasitoids of insect eggs. Trichogrammatids are among the smallest insects, ranging in size from 0.2 – 1.5 mm. The family is represented by over 800 described species in approximately 90 genera worldwide and is known from all vegetated terrestrial habitats (Lin 1994, Pinto 1997, Pinto 2006). The largest genera are *Trichogramma* and *Oligosita*, with ca. 180 and 95 species respectively (Noyes 2002, Pinto and Viggiani 2004). A broad range of insect hosts in several orders is known to be attacked, most prominently Coleoptera, Hemiptera, and Lepidoptera (Pinto and Stouthamer 1994). The most complete resource for the worldwide recognition and diagnosis of trichogrammatid genera is Doutt and Viggiani (1968), though the more recent publication by Pinto (2006) provides a comprehensive treatment of Nearctic taxa and a general discussion of the family. Other taxonomic works include a checklist of genera and species by Lin (1994), keys to Palearctic genera (Nikol'skaya 1952, Nikol'skaya and Trjapitzyn 1987), Indian genera (Yousuf and Shafee 1987), and to Nearctic genera (Pinto 1997). Family biology is reviewed in Pinto and Stouthamer (1994) and Pinto (1997).

Partly due to their small size and soft-bodied nature, and the consequent need for specialized collecting techniques, Trichogrammatidae have been inadequately sampled throughout the world and collections required for comprehensive taxonomic studies do not yet exist. It is clear that we currently know but a fraction of the true diversity of the family, and conservative estimates indicate that there may be more than 4000 additional species to be described (J. Pinto, unpublished). Results of year round collecting with Malaise traps in southern California suggest that trichogrammatids represent 10% of the local chalcidoid fauna (J. Pinto, unpublished). However, most material cannot be identified to species, and in some cases cannot even be accurately ascribed to any current genus. The confusion still present in the Trichogrammatidae can only be alleviated with further collecting efforts and the necessary taxonomic work.

Relatively few revisionary studies of trichogrammatids have been published. No large genus has been revised on a world-wide basis. Those generic treatments that have appeared either have limited geographic scope or deal with genera with few nominal species. Groups recently treated include *Paratrichogramma* (Doutt 1973), *Trichogrammatoidea* (Nagaraja 1978), *Soikiella* (Velten and Pinto 1990), *Xiphogramma* (Pinto 1990), *Ceratogramma* (Pinto and Viggiani 1991), *Uscana* (Fursov 1994), *Mirufens* (Neto and Pintureau 1997), *Trichogramma* (Pinto 1999), *Adryas* (Pinto and Owen 2004) and *Kyuwia* (Pinto and George 2004). The consequence of limited revisionary work is that most trichogrammatids cannot be placed to species. This is not only because many remain to be described, but also because described species are often difficult to identify due to the inadequacy of species descriptions and type material.

The taxonomy of the Trichogrammatidae is still in its infancy, and will probably continue to evolve as increased effort is focused upon the group (e.g. Pinto 2006, Owen et al. 2007). The currently most commonly followed classification is that of Viggiani (1971), which is based on male genitalia. This classification recognized two subfamilies, Trichogrammatinae and Oligositinae, based on varying levels of genitalic simplification. Although the placement of a proposed tribe and several other genera has been debated (cf. Pinto 2006, Owen et al. 2007), this basic structure for the family has been corroborated with independent molecular data (Owen et al. 2007). All evidence suggests that *Ufens* is a member of the Oligositinae (Pinto 2006, Owen et al. 2007), though sister groups are undetermined (see discussion below).

Biological Control. According to Noyes (1985) Trichogrammatidae ranks as the seventh most successful hymenopteran family utilized in biological control. This high ranking is due largely to *Trichogramma*, used extensively in applied entomology because its members parasitize numerous lepidopteran pests and it can be mass propagated and released with relative ease. *Trichogramma* is the world's most widely used arthropod for augmentative biological control programs (Smith 1996), and is a potentially effective biological control agent in a wide range of systems (Li 1994). Use of trichogrammatids as biological control agents, however, is not restricted to *Trichogramma*. Among others, the association of *Ufens* with the glassy-winged sharpshooter, *Homalodisca coagulata* (Say) and the smoketree sharpshooter, *Homalodisca liturata* Ball (= *H. lacerta* (Fowler)) (Hemiptera: Cicadellidae: Proconiini) has been particularly well documented (Triapitsyn 2003, Al-Wahaibi et al. 2005). These sharpshooters are important vectors of the bacterium *Xylella fastidiosa,* which causes diseases on several crops and ornamentals including Pierce's disease of grapes, phony peach disease, almond leaf scorch, alfalfa dwarf, and oleander leaf scorch (Blua *et al.* 1999, Varela *et al.* 2001). Powers (1973) reported an *Ufens* sp. attacking *H. liturata* on *Hibiscus syriacus* L. (Malvaceae). More recent observations suggest that egg masses of *Homalodisca* on *Simmondsia chinensis* (Link) Schneid. (Simmondsiaceae) are predominantly parasitized by two *Ufens* species (Al-Wahaibi 2004). These *Ufens* parasitoids are also shown to be responsible for a large proportion of *Homalodisca* egg parasitism on cultivated plants, such as citrus, in Riverside, California (Al-Wahaibi 2004). There has been interest in using *Ufens* species as part of a biological control effort against *H. coagulata* (Triapitsyn and Hoddle 2001, 2002, Triapitsyn *et al.* 1998, 2002). This effort was hampered by difficulties in rearing *Ufens* species in quarantine, leading to the hypothesis that *Ufens* species might be hyperparasitoids, attacking the primary parasitoids, *Gonatocerus* species (Hymenoptera: Mymaridae), inside *Homalodisca* eggs (Triapitsyn, 2003). This hypothesis was contradicted by Al-Wahaibi et al. (2005), who found *Ufens* to be primary parasitoids. These wasps may be challenging to rear because they parasitize host eggs only immediately after being laid, a condition usually not met using standard parasitoid rearing protocols.

Ufens Revision. This work represents the first attempt to completely revise *Ufens*, and thus suffers from several limitations. For example, relatively few geographical regions have been reasonably well sampled (e.g. United States; Australia: Queensland), while broad geographical areas (e.g. Africa) remain poorly collected for microhymenoptera. Attempts were made to accumulate specimens from throughout the world, but clearly more collecting is needed to appreciate *Ufens* diversity. In particular, the diversity of the few collections from Africa and southeastern Asia indicate that these areas are likely to harbor further undescribed species. Although Australia has been well collected in comparison to many other places, the number of species represented there by one or a few specimens indicates that this continent is also likely to hold additional undescribed species.

The problem of few specimens per species is common. Although approximately 2,000 specimens were examined, most species are represented by relatively small series. While most *Ufens* are readily diagnosable by the morphology of male genitalia, small series generally result in an underappreciation of intraspecific morphological diversity and hinder recognition of possible cryptic species complexes. As shown by Pinto et al. (1989) for *Trichogramma*, there is the potential for significant phenotypic plasticity in minor anatomical characters within a species. Similarly, distinct species of *Trichogramma* are known to differ by minute anatomical distinctions (Pinto et al. 1986, 1997), or none at all (Pinto et al. 2003). Obviously, neither intraspecific variation nor minor species differences can be adequately appreciated if very few specimens are available for study. Molecular diversity was also difficult to ascertain, as DNA sampling was possible for only 12 of the 43 species included in this revision. These limitations suggest that the 43 species treated here are probably a considerable underestimate of the true diversity of *Ufens*, and that any attempt to infer species distributions and phylogenetic relationships are also extremely preliminary.

As found in *Trichogramma* (Pinto 1999), male genitalia of *Ufens* appear to be the most efficient morphological character system for species separation. Unfortunately, this approach prohibits the use of females for many of the species studied, as positive associations between males and females are difficult without rearing records. This problem is particularly acute for specimens from areas with apparent high levels of sympatry, such as Australia. This study therefore relies heavily on male genitalia for identification and classification of species. Consequently, females remain unknown for many species described in this revision. Due to this dependence on male genitalia, several *Ufens* species previously described only from females cannot be incorporated into the current framework (Table 2). It is likely that certain of these *nomina dubia* will never be identifiable, but it is hoped that correct associations can be made once type localities become better sampled.

The distinctiveness of *Ufens* male genitalia contrasts with the relative uniformity of other morphological traits. The non-genitalic characters utilized in this study were primarily taken from antennal morphology, mesosomal sculpturing, and wing setation. Unfortunately, many of the non-genitalic characters diagnosed for *Trichogramma* (Pinto 1999), were not found to be useful for *Ufens*, either due to too much or too little variation within and among species.

Many of the limitations noted above are not unique to *Ufens* but are shared with numerous other chalcidoid taxa. Shortcomings notwithstanding, a world-wide revision of *Ufens* is timely. Thanks to a recent sampling effort by parasitic hymenopterists throughout the world a far greater amount of material than ever before is now available. The approximately 2,000 individuals of *Ufens* used for this study, including 1,100 slide-mounted specimens, represent a dramatic increase in material available. In addition, unlike the situation in certain taxa, most of the type specimens were readily available from museums such as National Museum of Natural History (Washington, DC), Queensland Museum (Brisbane), and the Natural History Museum (London). Furthermore, the considerable interest in *Ufens* for biological control of *Homalodisca* species and other pests (Triapitsyn and Hoddle 2001, 2002; Triapitsyn *et al.* 2002) requires accurate identifications, which only detailed revisionary work can provide. It is hoped the current work will facilitate further studies on this genus that will provide greater insight into diversity, phylogenetic relationships, and the role its species play in the ecosystems where they are found.

Materials and Methods

Specimen Preparation and Examination. Slide-mounted specimens were prepared in Canada balsam as described by Platner et al. (1999), with the antenna, fore leg, fore wing and hind wing generally removed from the right side of the body and mounted under a separate coverslip. Slides were prepared using a Zeiss Stemi SV 6 dissecting microscope with a Diagnostic Instruments illuminated base. They were examined and measured at magnifications of up to 600x using a Zeiss Axioscope 2 compound microscope, with the measurements calibrated using an eyepiece micrometer with 0.01mm divisions. Card-mounted specimens were prepared using Hexamethyldisilazane (HMDS) (Heraty and Hawks 1998), though some specimens not personally prepared may have been critical point dried (Gordh and Hall 1979). Card mounted specimens were prepared and examined with the Zeiss Stemi SV 6 microscope. Scanning electron microscope (SEM) images were taken with a Phillips XL30-FEG. Specimens utilized were dried using HMDS (Heraty and Hawks 1998) and dissected under the Zeiss Stemi SV 6 microscope using a tool made with a 0.01 inch diameter minuten pin. Body parts were mounted on a 12.7 X 3.2 mm Leica/Cambridge aluminum SEM stubs with double-sided carbon tape. Mounted specimens were sputter coated using an Emscope ES500 with a gold-palladium mixture.

Digital images of slide- and card-mounted material were prepared using the Automontage image capture system (Microbiology International, Synchroscopy). Images were captured using a JVC KYF-70 color video camera mounted on a Zeiss Axioscope 2 compound microscope or a Leica WILD M10 compound microscope. SEM and Automontage images were cropped and adjusted for contrast and brightness using Adobe Photoshop 8.0.

Terminology and Measurements. Terminology is based primarily on Gibson (1997), although some features, especially those relating to genitalia, wings and antennae, are based on Doutt and Viggiani (1968), Viggiani (1971), Amornsak et al. (1998), and Pinto (1999).

Body
Size and color: **Body length (BL)** is the maximum length from the anterior margin of the pronotum to the posterior margin of the last gastral tergum. **Color** was determined using card-mounted specimens which had been curated using HMDS whenever possible, or estimated from slide-mounted material. As in *Trichogramma* (Pinto et al. 1989), color is generally of limited use in *Ufens*. Color of the various species ranges from almost entirely yellow to primarily dark brown. In one case color is sexually dimorphic, but this is exceptional.

Head
Antenna: Sections of the antennal flagellum include the **anellus (A)**, **funicle (F)** and **club (C)** (also known as clava); specific segments of these sections are indicated by numbers following the acronym (e.g. C2 = 2nd club segment) (Fig. 6).

Maximum club length is compared with funicle length (C/F). The maximum length of F2 is also compared to the maximum length of F1 (F2/F1). Terms for antennal sensilla and setation are derived from those used for *Trichogramma*. Comparisons are based on an examination of all species with a light microscope and of some species with the scanning electron microscope. Terms follow Vincent and Goodpasture (1986) [=V/G], Olson and Andow (1993) [=O/A] and Pinto (1999) [=P]. The seven recognized types of sensilla (cf. Fig. 6) are: 1) **Aporous sensilla trichodea B (APB)** [O/A]. APB are short, socketed, found on the pedicel and usually on the funicle of females; they are uncommon on the funicle of males. 2) **Placoid sensilla (PLS)** [P]. PLS are elongate and spatuliform, and are generally arranged longitudinally on an antennal segment. They are typically found on all segments of the funicle and club, except for the terminal club segment in males. 3) **Basiconic peg sensilla (BPS)** [P]. BPS are bulbous structures generally occurring at the apex of each funicular and club segment, except the C4 of males. 4) **Flagelliform setae (FS)** [V/G, P] (also known as multiporous pitted sensilla trichodea A [O/A]). FS are elongate, slightly curved and presumed to be unsocketed, though there is a depression at their base which could be interpreted as a socket. They form the setal "whorls" characteristic of *Ufens* male antennae (Fig. 6). 5) **Unsocketed setae (US)** [P] (also known as aporous sensilla trichodea A). US are short, stiff, procumbent hairs; they are considerably shorter than FS, and clearly lack a basal socket. They are more common on female than on male antennae. 6) **Uniporous pit pore sensilla trichodea D (UPP)** [O/A]. UPP are relatively short, apically curved, socketed sensilla; a single UPP occurs at the apex of C3 in females. 7) **Coeloconic sensilla (CS)** [V/G]. CS are pit organs with a small protruding peg. They are found on A2 and F2 only, but can be very difficult to discern.

Mesosoma

Sculpturing. Descriptions of sculpturing were based on an examination of the midlobe of the mesoscutum anterolaterally. Sculpturing tends to be more tightly compressed medially and therefore more homogenous between species. This trait has generally been ignored in trichogrammatids (e.g. Pinto 1999), but in *Ufens* it is sufficiently species-specific to warrant attention. Terminology used is modified from Harris (1979). *Ufens* and outgroups have sculpturing which is either **striate** or **cellulate**, with varying degrees of interstitial sculpturing between striae or within cells (cf. Figs. 18, 39, 50).

Hind tibia. **Hind tibial length (HTL)** is the maximum length of the hind tibia. HTL is commonly used in various ratios, including BL/HTL.

Wings. Wing characters can be important for distinguishing some species, but generally only when combined with other traits. **Fore wing length (FWL)** is measured from the apex of the humeral plate to the wing apex; **fore wing width (FWW)** is measured at the widest point of the wing, in its apical third; their ratio, FWL/FWW, is reported. **Fore wing fringe seta length (FWFS)** is measured where longest at the posterolateral margin of the wing and is reported in relation to FWW. **Hind wing length (HWL)** is measured from the base of venation to apex of wing, while **hind wing width (HWW)** is measured immediately apical of the hamuli; their ratio, HWL/HWW, is reported. **Hind wing fringe seta length (HWFS)** is measured where longest, generally in the apical third of the posterior margin of the wing and is reported in relation to HWW. Fore wing veins considered include the **submarginal (SM)**, **premarginal (PM)**, **marginal (MV)** and **stigmal (SV) veins** (Fig. 7). The **radial process (RP)** refers to a sclerotized spur radiating from the base of the PM. Ratios used include MV length compared to PM length (MV/PM), and SV length compared to MV length (SV/MV). MV length is also related to its width (MV length/MV width). The SM and PM veins, and to a lesser extent the MV, represent the posterior limits of the **costal cell (CC)**. All the major setal tracks in the trichogrammatid fore wing disk as outlined by Doutt and Viggiani (1968) are present in *Ufens*. These include the **radius (R)**, **radial sector 1 (RS_1)**, **radial sector 2 (RS_2)**, **median vein track (M)**, **r-m crossvein track (r-m)**, **cubital track 1 (CU_1)**, **cubital track 2 (CU_2)**, **anal vein track (A)**, and **basal vein track (B)** (Fig 7). The **maximum distance from r-m to M (Max r-m to M)**, measured near the wing apex, is compared to the **minimum distance (Min r-m to M)** between these two tracks, measured at approximately mid-disk. Minute cuticular nub-like projections on the ventral surface of the fore wing of some species are referred to as **alar acanthae (AA)**. **Setal density** on the fore wings varies both interspecifically and, usually less extensively, intraspecifically. Intraspecific variation appears to be correlated with body size. Setal density can be approximated by determining the number of setae between RS_2 and r-m. It is regarded as light in species with less than 110 setae in this area, and as heavy in species with more than this number.

Metasoma
The only characters of the metasoma utilized in *Ufens* taxonomy are the ovipositor and male genitalia.

Ovipositor. **Ovipositor length (OL)** can vary considerably interspecifically (Fig. 9). Determining levels of intraspecific variation in length was not straightforward in many cases because of the difficulty of recognizing conspecific females without associated males. Ovipositor length is presented as a ratio to hind tibial length (OL/HTL). No other character of the ovipositor was found to be of taxonomic value.

Male genitalia. Characteristics of the male genitalia, such as overall shape and presence and absence of certain structures, are the most critical features for identification of *Ufens* species. In fact, most species can be identified using male genitalia alone. The magnitude of genitalic diversity, however, renders determination of homology across all species difficult. Nevertheless, many of the same generalized structures found in the Trichogrammatidae by Viggiani (1971), and reiterated for *Trichogramma* (Pinto 1999), can be identified in *Ufens* males. Unlike *Trichogramma*, however, the aedeagus of *Ufens* is indistinct from the genital capsule. Due to the difficulties of comparisons between species, the generalized male genitalia of *Trichogramma* are presented as a model against which *Ufens* species are compared (Fig. 8).

Genital capsule length (GL) and **width (GW)** represent maximum values. GL is compared to both GW (GL/GW) and HTL (GL/HTL). The genital capsule of some species shows an **anterior invagination (AI)** (e.g. Figs. 30, 44). The maximum depth of the invagination relative to overall capsule length is reported (AI/GL). The opening on the dorsum of the genital capsule through which the sperm duct enters is the **anterodorsal aperture (ADA)** (Fig. 8), and its length is also compared to the GL (ADA/GL). **Aedeagal apodemes (AP)** are present only in a handful of *Ufens*, and are considered homologous with the anterior extensions of the aedeagus in other trichogrammatids (e.g. Figs. 17, 44). Their length is compared to GL (AP/GL) when applicable. **Parameres (PAR)** are usually present, and are generally the apicolateral-most structures of the genitalia. They tend to arise from the ventral surface, and are therefore obscured in some species when examining the genitalia dorsally. In *Ufens* the parameres usually possess a distinctive terminal spine at their apex (e.g. Figs. 18, 52). It is unclear if this structure is socketed or not, but it is clearly differentiated from the apex of the paramere due to its reduced width. However, several species have parameres that lack a distinct terminal spine (e.g. Figs. 20, 32). When present, PAR length is compared with GL (PAR/GL). The overall shape and point of origin of PAR relative to the ADA can be important taxonomic features. **Volsellae (VS)** are sometimes present in *Ufens* male genitalia, though they can be difficult to identify without use of the scanning electron microscope. They are generally located medial to the parameres (e.g. Figs. 20, 48), and their length is also compared to GL (VS/GL) when possible. The **ventral process (VP)** is another structure apparently present in some species and absent in others (e.g. Figs. 25, 52). When present its length is compared with GL (VP/GL). It occurs ventrally along the midline of the genital capsule, and it is generally widest at the base and attenuate apically, though its shape can vary from spinose to broadly subtriangular. This structure is known as the intervolsellar process in some studies (e.g. Pinto 1999), but the term ventral process is preferred here as volsellae are absent or difficult to distinguish in many *Ufens* species. A **dorsal ridge (DR)** is present in some species, and is identified as a raised or more heavily sclerotized line running along the midline on the dorsal surface of the anteroventral area of the genital capsule (e.g. Figs. 37, 45). Some *Ufens* species possess a **dorsal projection**, which may or may not be homologous with the dorsal lamina of *Trichogramma*, a structure considered unique to that genus (Pinto 1999). This structure arises from the dorsal surface immediately posterior to the termination of the ADA, and is somewhat flap-like as it extends posteriorly to the end of the genital capsule (e.g. Figs. 45, 52). The dorsal projection can be difficult to see in slide-mounted individuals and is most easily appreciated in SEM micrographs. It seems likely that this structure is fused to the apical portion of the genital capsule in those species which do not possess it. A transverse line, called the **transverse hinge,** appears across the posterior portion of the genital capsule in some species (e.g. Figs. 20, 25). Although its function as a hinge has not been verified, the genitalia of some individuals of species possessing the transverse hinge have been observed bent at this point (e.g. Fig. 20), lending credence to the hypothesis that this area acts as a hinge point during copulation.

Description Format. Species treatments follow a description of *Ufens*. Characters that do not vary within the genus are not discussed in the individual species descriptions. Descriptions are based on all available material. Previously described *Ufens* species are redescribed. In certain cases only the original descriptions and type specimens were available. As noted, in other cases, additional material not in the type series but identified as conspecific was incorporated in the redescription. Valid species are presented in alphabetical order, followed by a section of *nomina dubia*. Acronyms for institutions follow the Bishop Museum checklist (http://hbs.bishopmuseum.org/codens/codens-inst.html).
The following institutional acronyms are utilized herein:
ANIC – Australian National Insect Collection (CSIRO), Canberra, Australia
BMNH - British Museum of Natural History (Natural History Museum), London, U.K.
BPBM - Bernice P. Bishop Museum, Honolulu, Hawaii, U.S.A.
CNC - Canadian National Collection of Insects, Ottawa, Ontario, Canada
DEZA - Dipartimento di Entomologia e Zoologia Agraria dell'Università, Portici, Italy
FACS - Fujian Agricultural and Forestry University, Fuzhou, China
UCRC – University of California Research Collection, Riverside, California, U.S.A.
USNM - National Museum of Natural History (NMNH), Washington D. C., U.S.A.
QM – Queensland Museum, Brisbane, Australia
ZMUC - University of Copenhagen, Zoological Museum, Copenhagen, Denmark

The following sections are included in most species treatments: diagnosis, type information, etymology, distribution, biological information, description, other material examined and comments. The species *diagnosis* includes traits considered to be particularly diagnostic that should aid in differentiating it from other species. It also contains some generalized information about traits that may be common to multiple species. It is hoped that this information helps in the recognition of females, as females are unknown for many of the species. The *type specimens* section includes all pertinent data for type specimens, including their deposition. All types or type series indicated with a '♦' have been examined, unless otherwise stated for each species. For new species, an allotype and paratypes are not always designated. Due to the potential difficulty of associating individuals and the possibility of cryptic species, type series were restricted to material collected from the same locality, or preferably, reared from the same host. Label information listed in quotation marks " " is written exactly as it appears on specimen labels. The *etymology* section explains the origin of newly proposed specific names and, in the generic redescription, conjecture about the etymology not explicitly explained when originally proposed. The *distribution* section summarizes known geographical distribution. The *biology* section summarizes known hosts and host plants. *Species descriptions* contain considerable quantitative data. All measurements are given as means followed by sample ranges. In species with few individuals, sample ranges approach or equal observed ranges. Unless otherwise indicated, n = 5 for all measurements. Due to the paucity of identifiable females in many species, males were the primary sex measured for all but female-specific characters. No differences were observed in non-sexually dimorphic characters such as wings and sculpturing. An attempt was made to represent both the size and geographic variability of the species. A list of all additional (non-type) *material examined* is given. Finally, the *comments* section provides additional notes about the species, characteristics of type material, identification in the literature, or any other information deemed pertinent. This section is omitted in certain cases when no additional information is necessary.

Institutions housing material examined other than types are indicated only when n ≤ 5. Any comments interjected by the author appear in brackets []. Collection method abbreviations found in material examined are as follows: sweep = SW, yellow pan trap = YPT, and Malaise trap = MT. Locality information commonly abbreviated include: County = Co., National Park = N. P., National Forest = N. F., mountain = Mtn, University of California, Riverside = UCR. All dates are presented in the form 'day.month (roman numeral).year' (e.g. 12.xi.2002). Abbreviations for frequently cited collectors include: Jeremiah George = JG, John M. Heraty = JMH, James B. Munro = JBM, Albert K. Owen = AKO, John D. Pinto = JDP, Gary Platner = GP, James B. Woolley = JBW. Other common abbreviations include emerged =em., elevation = el., miles = mi., kilometer = km

Phylogenetic Relationships.
Analyses. To estimate phylogenetic relationships of *Ufens* species, maximum parsimony (MP) analyses were performed using PAUP*4.0β10 (Swofford 2001). Two versions of the data set were utilized, one with all 43 ingroup taxa and 5 outgroup taxa, and a reduced data set including only those taxa for which both molecular and morphological information were available, i.e. 12 ingroup and 4 outgroup taxa. Morphological analyses consisted of 37 characters with ca. 4% missing/inconclusive data. Molecular analyses were performed with 1423 characters from rRNA 28S D2 and D3 aligned according to secondary structure (cf. Owen et al. 2007). The analyses emphasized here include: a) combined molecular and morphological for only those taxa with both types of data (Fig. 1), b) only morphological data for all taxa (Fig. 2-3), and c) morphological data using the result of the molecular and morphological analysis as a backbone (Fig. 4) (see below). Analysis of the reduced, combined molecular and morphological, data set was performed with branch and bound keeping only minimum length trees. All other analyses were performed with 500 random stepwise heuristic searches with tree bisection reconnection (TBR) branch swapping, saving a maximum of 100 trees per repetition. All characters were treated as unordered and with equal weights. Gaps in the molecular alignment were treated as missing data. Bootstrapping was performed with 500 replicates of 5 random addition-sequences each, saving a maximum of 100 trees. All sets of trees were condensed and filtered using best score after MP analysis. After initial analyses, characters were weighted through successive

approximations (SAW) according to their rescaled consistency indices using a base weight of 1000 until their tree length was stable (Farris 1969). Resulting trees were rescaled to unity for comparison to the original set of MP trees (Babcock et al. 2001). Strict consensus trees were generated from both results of unweighted and SAW analyses, though SAW results are primarily presented here. Morphological analyses were performed leaving the ingroup both unconstrained (Fig. 2) and constrained (Fig. 3) as monophyletic. Due to having greater confidence in the results of the reduced data set (morphological + molecular), the result from the strict consensus of this analysis was utilized as a 'backbone' for further iterations of all taxa utilizing only morphological data (Fig. 4). Use of a backbone constraint tree maintains overall structure of the most resolved topology (strict consensus in this case), but allows for the addition of new taxa as long as the relationships among the original taxa remain the same (Swofford 2001).

Outgroups. Five outgroup taxa are included. No definitive sister group has been proposed for *Ufens* based on either morphological or molecular means, making a choice of outgroups problematic for this study. Outgroups were chosen based on the following criteria, and using molecular results (Owen et al. 2007) as a guideline: 1) basal and primitive (*Ceratogramma*); 2) previously considered closely related to *Ufens*, but most likely basal (*Brachyufens* (Doutt and Viggiani 1968) and *Mirufens* (Yousuf and Shafee 1987)); 3) potentially closely related based on at least some of the molecular results (*Monorthochaeta*). The only outgroup taxon lacking molecular data is an undescribed genus from Botswana (New Genus Botswana). It is similar to *Mirufens* in its male antenna, male genitalia, and long, sweeping stigmal vein, but lacks its 2-segmented maxillary palps and transversely ridged pedicel. This new genus is included in analyses as certain characters indicate ties to *Ufens*, although its male genitalia clearly place it as a member of the Trichogrammatinae.

Characters. The following characters were utilized in phylogenetic analyses. The complete character matrix can be found in Table 1. Unless otherwise indicated, figures mentioned are meant to be exemplary, but do not necessarily encompass all known variation of each character.

Fore wing (1-5) (cf. Fig. 7)
1. Alar acanthae: (0) present; (1) absent. The polarity of this character is in question, as it appears to vary both within the outgroup assemblage and among *Ufens* species. Although numbers of alar acanthae vary intraspecifically, their presence or absence appears to be stable within species.
2. Maximum/minimum distance from r-m to M ratio: (0) < 3; (1) >3. The setal tracks r-m and M do not diverge drastically in most species, but they do in some (e.g. *U. niger*, Fig. 39c; *U. similis*, Fig. 47b). Several species (e.g. *U. principalis*, Fig. 45c; *U. simplipenis*, Fig. 48c) exhibit ratios that span both character states and are coded as polymorphic.
3. Area between setal tracks CU1 and CU2: (0) with more than one track or with numerous dispersed setae; (1) with a single setal track. Some species coded as '1'

may sometimes have a couple of extra setae not confluent with the single track. However, taxa coded as '0" clearly have many extra dispersed setae, which appears to be indicative of an overall densely setose fore wing. Most outgroup members are coded '0', whereas most *Ufens* species are coded '1'.

4. RS1 setal track: (0) absent or indistinguishable from surrounding setae; (1) present. This setal track is distinguishable in all taxa except outgroup taxa *Ceratogramma masneri* and *Monorthochaeta nigra*.

5. Costal cell: (0) with ≤ 1 complete setal track; (1) with 2 distinct setal tracks. All outgroups except *Brachyufens* have, at most, 1 complete setal track within the costal cell. Most *Ufens* have 2 distinct tracks of varying numbers of seta. This character may be related to the size of the costal cell itself, which is fairly large in *Ufens* and smaller in outgroup taxa.

Hind wing (6-7)

6. Discal width: (0) decreases immediately beyond hamuli (e.g. *U. niger*, Fig. 39d; *U. similis*, Fig. 47c); (1) does not decrease or actually increases immediately beyond hamuli (e.g. *U. debachi*, Fig. 20d; *U. hercules*, Fig. 29c). All outgroups and many ingroup taxa possess hind wings that begin to narrow immediately apical of the hamuli.

7. Discal setation: (0) consisting of many dispersed setae; (1) consisting of only 3 distinct setal tracks and very few additional setae. Only the outgroup taxa, *Brachyufens* and *Ceratogramma*, exhibit state '0'. Although their dispersed setae sometimes align to form tracks, the majority of discal setae are not associated with those tracks.

Head and Male Antenna (8-21) (cf. Fig. 6)

8. Maxillary palps: (0) 2-segmented; (1) 1-segmented. There appears to be a trend in the Trichogrammatidae towards a reduction in palpal segments in the more derived elements of the family including *Ufens* and other genera, and the only representatives in analyses with 2-segmented palps are the outgroup taxa *Ceratogramma*, *Brachyufens* and *Mirufens*.

9. Club segment number in male: (0) ≤ 3; (1) 4. A small, terminal C4 is characteristic of *Ufens*, but also is found in *Mirufens* and New Genus Botswana.

10. Size of C4: (0) normal, easily distinguished from C3; (1) minute, difficult to distinguish from C3 and not extending beyond apex of terminal PLS of C3. The small C4 is a unique trait of *Ufens messapus* (Fig. 35a) and *U. spicifer* (Fig. 49a).

11. Aporous sensillar trichodea B (APB) sensilla on funicle: (0) absent; (1) present. While APB are present on the funicle of many trichogrammatids, including *Ufens* females, they are rarely present on the funicle of *Ufens* males.

12. Unsocketed seta (US) on funicle: (0) absent; (1) present. US are generally not present on the funicle of *Ufens* males.

13. Flagelliform seta (FS) on F1: (0) present; (1) absent. FS are usually present on F1, except for a few *Ufens* species and the outgroup *Ceratogramma* and *Monorthochaeta*.

14. Flagelliform seta (FS) arrangement on funicle and/or club: (0) not arranged in a whorl; (1) arranged in a whorl. The almost linear arrangement of FS around the funicle and club of *Ufens* males is characteristic, and is shared by New Genus Botswana.

15. Placoid sensilla (PLS) on funicle: (0) absent; (1) present. PLS are always present on the funicle of male *Ufens*, and absent in all outgroups except *Monorthochaeta* and New Genus Botswana.

16. Placoid sensilla (PLS) on F1: (0) ≤ 1; (1) >1. Only a few *Ufens* species have more than 1 PLS on F1 (e.g. *U. gloriosus*, Fig. 28a), and none of the outgroups has.

17. Placoid sensilla (PLS) on F2: (0) ≤ 1; (1) >1. As in trait 16, only a few *Ufens* species have more than 1 PLS on F2, and none of the outgroups has. Although most taxa with additional PLS on F1 also tend to have them on F2, this does not hold for all species (e.g. *U. aperserratus*, Fig. 12a), suggesting that these traits are independent.

18. Placoid sensilla (PLS) on C1: (0) ≤ 1; (1) >1. All taxa except *U. gloriosus* (Fig. 28a), *U. parvimalis* (Fig. 42a) and *U. placoides* (Fig. 44a) have 1 or 0 PLS on C1.

19. Placoid sensilla (PLS) on C2: (0) ≤ 1; (1) > 1. A few *Ufens* species (e.g. *U. gloriosus*, Fig. 28a) and the outgroup taxa *Ceratogramma* and *Monorthochaeta* have more than 1 PLS on C2.

20. Placoid sensilla (PLS) on C3: (0) ≤ 3; (1) > 3. A few *Ufens* species (e.g. *U. gloriosus*, Fig. 28a) have more than 3 PLS on C3, and none of the outgroups do. As with characters 16 and 17, characters 18-20 are believed to be independent.

21. Club segments: (0) not separated by a deep constriction; (1) separated by a deep constriction. There is variation in the compactness of club segments, but several species have dramatically obvious constrictions (especially anteriorly) between segments (e.g. *U. nazgul*, Fig. 38a).

Male Genitalia (22-37) (cf. Fig. 8)

22. Aedeagus: (0) distinct from genital capsule; (1) indistinct from capsule. All outgroups except *Monorthochaeta* have an aedeagus separate from the capsule. Corresponding with the trend towards fusion of the aedeagus and capsule in the more derived elements of the family, the aedeagus of all *Ufens* is indistinct from the capsule.

23. Genitalia shape: (0) not bulbous near posterior end; (1) bulbous posteriorly. *U. austini* (Fig. 14d) and *U. nazgul* (Fig. 38d) have a distinctly bulbous area immediately posterior of the transverse hinge.

24. Basal margin invagination of genital capsule: (0) absent or slight (invagination <0.2 genitalic length); (1) deep (invagination >0.2 genitalic length). Only *U. ceratus* (Fig. 18g-i) and *U. invaginatus* (Fig. 30d) have a dramatically invaginated genital capsule. A further division of state '0' was considered, as the capsule in some species has no demonstrable invagination and in others it is consistently slightly invaginated. However, objective coding of these additional character states is precluded by considerable overlap among species.

25. Dorsal ridge: (0) present (e.g. *U. similis*, Fig. 47d); (1) absent. *Ceratogramma* and *Monorthochaeta* are the only outgroup genera without a dorsal ridge, and it is a variable character within *Ufens*.

26. Maximum width of anterodorsal aperture (ADA) relative to genital capsule width: (0) nearly as wide (e.g. *U. principalis*, Fig. 45f-g); (1) distinctly narrower (e.g. *U. rimatus*, Fig. 46d). Only 4 species of *Ufens*, and none of the outgroups, have an ADA which is distinctly narrower than the width of the genital capsule. In most species, the maximum capsule width is at the ADA.

27. Anterodorsal aperture (ADA) shape (outline in dorsal view): (0) uniform its entire length; (1) distinctly constricted and narrowed in posterior half (nearly spatulate). The ADA of most *Ufens* species and all outgroups has a fairly uniform curvature, whereas several species (e.g. *U. thylacinus*, Fig. 51d) show an abrupt constriction, then nearly parallel sides anteroposteriorly.

28. Ventral process: (0) absent; (1) present. *Ceratogramma* and *Monorthochaeta* are the only outgroup members possessing a VP. It is present in many *Ufens* species, though its shape can vary considerably. Recognition of volsellae is sometimes difficult due to the diversity of *Ufens* genitalia. Their identification was made according to relative position on the genitalia and by comparisons to the generalized trichogrammatid model (Fig. 8).

29. Ventral process: (0) entire; (1) bifid or with lateral spine. The ventral process of most *Ufens* and all outgroups possessing this appendage is simple and unbranched. However, the ventral process of *U. kender* (Fig. 31f) has a small spine in its apical third, and that of *U. nazgul* (Fig. 38d) and *U. placoides* (Fig. 44d) is asymmetrically bifurcate, with one side of the bifurcation long and approaching the apex of aedeagus.

30. Ventral process base: (0) narrow, maximum width < half width of capsule at base of process (e.g. *U. flavipes*, Fig. 25f); (1) present, wide, maximum width > half width of capsule at base of process (e.g. *U. principalis*, Fig. 45h).

31. Transverse hinge: (0) absent; (1) present. The transverse hinge, apparently a novel structure within the Trichogrammatidae, is shared by a number of *Ufens* species (e.g. *U. debachi*, Fig. 20g; *U. flavipes*, Fig. 25e).

32. Aedeagal apodemes: (0) present; (1) absent. Apodemes are present in all outgroup taxa but only in a few *Ufens* species (e.g. *U. decipiens*, Fig. 21d; *U. placoides*, Fig. 44d).

33. Volsellae: (0) present (e.g. *U. ceratus*, Fig. 18i); (1) absent (e.g. *U. flavipes*, Fig. 25e-f). Volsellae are present in all outgroups, but are apparently variable within *Ufens*. This may be due in part to the difficulty of detecting them in slide-mounted specimens of some species.

34. Parameres: (0) present (e.g. *U. ceratus*, Fig. 18i); (1) absent (e.g. *U. flavipes*, Fig. 25e-f). Most *Ufens* species and all outgroups possess parameres. Recognition of parameres is sometimes difficult due to the strong modification of many *Ufens* genitalia. Identification was based on relative position and comparisons to the generalized trichogrammatid model (Fig. 8).

35. Terminal spine of parameres: (0) present (e.g. *U. ceratus*, Fig. 18i); (1) absent (e.g. *U. debachi*, Fig. 20g). The parameres of all outgroup taxa possess a terminal spine, as they do in a large number of other trichogrammatid genera. Their presence varies within *Ufens*.

36. Paramere width: (0) subequal in width their entire length (excluding terminal spine) (e.g. *U. ceratus*, Fig. 18g-j); (1) distinctly wider at base (e.g. *U. pallidus*, Fig. 41d); (2) wider near middle or apex (e.g. *U. lanna*, Fig. 34d). Parameres are subequal in width their entire length in most taxa, including all outgroups. However, a number of species demonstrate the alternative states.

37. Paramere base relative to posterior edge of ADA: (0) positioned distinctly anteriorly (e.g. *U. austini*, Fig. 14d); (1) positioned evenly (e.g. *U. principalis*, Fig. 45f, h-i); (2) positioned posteriorly (e.g. *U. simplipenis*, Fig. 48i-j). This is a variable character among *Ufens* species. It is also problematic to code for those outgroup taxa that do not have a distinct ADA, but rather a dorsal trough (*Brachyufens*, *Ceratogramma* and *Mirufens*) which is considered homologous to the ADA. Since the parameres of these taxa are inserted where the trough ends they are coded as state '1' for consistency. *Monorthochaeta* has an ADA but differentiation of the parameres is difficult; it is not coded for this character.

Table 1. Morphological matrix used in phylogenetic analysis of *Ufens*. Characters and states are discussed in text.

Taxon	1	2	3	4	5	6	7	8	9	10	11	12	13
Ceratogramma masneri	1	?	?	0	0	0	0	0	0	?	1	1	1
Mirufens sp.	0	0	0	1	0	0	1	0	1	0	0	0	0
Monorthochaeta nigra	0	?	?	0	0	0	1	1	0	?	1	1	1
Brachyufens osborni	1	0	0	1	1	0	0	0	0	?	0	0	0
New Genus Botswana	0	0	1	1	0	0/1	1	1	1	0	0	0	0
U. acacia	1	0	1	1	1	1	1	1	1	0	0	0	0
U. acuminatus	0	1	0	1	1	0	1	1	1	0	0	0	0
U. aperserratus	0	0	1	1	1	1	1	1	1	0	0	0	0
U. apollo	0	0/1	1	1	1	0	1	1	1	0	0	0	0
U. austini	0	0	1	1	1	1	1	1	1	0	0	0	0
U. australensis	0	0	1	1	1	1	1	1	1	0	0	0	0
U. bestiolis	0	0	1	1	1	1	1	1	1	0	0	0	0
U. cardalia	1	0	1	1	1	1	1	1	1	0	0	0	0
U. ceratus	1	0	1	1	1	1	1	1	1	0	0	0	0
U. cupuliformis	1	0	1	1	1	1	1	1	1	0	0	0	0
U. debachi	1	0	0/1	1	1	1	1	1	1	0	0	0	0
U. decipiens	0	0	1	1	1	1	1	1	1	0	1	1	0
U. dilativena	1	0	1	1	1	1	1	1	1	0	0	0	0
U. dolichopenis	0	1	0	1	1	0	1	1	1	0	0	0	0
U. elimaeae	0	0	1	1	1	0	1	1	1	0	0	1	0
U. flavipes	0	0	1	1	1	1	1	1	1	0	0	0	0
U. foersteri	0	0/1	1	1	1	0	1	1	1	0	0	0	0
U. forcipis	1	0	1	1	1	1	1	1	1	0	0	0	0
U. gloriosus	1	0	1	1	1	1	1	1	1	1	1	0	1
U. hercules	1	0	1	1	1	1	1	1	1	0	0	0	0
U. invaginatus	0	0	1	1	1	0	1	1	1	0	0	0	0
U. kender	0	0	1	1	1	1	1	1	1	0	0	1	0
U. khamai	1	0	1	1	1	1	1	1	1	0	0	0	0
U. kurrajong	0	0	1	1	1	0	1	1	1	0	0	0	0
U. lanna	1	0	1	1	0	1	1	1	1	0	0	0	0
U. messapus	0	0	1	1	1	0	1	1	1	1	1	1	1
U. mezentius	1	0	1	1	1	1	1	1	1	0	0	0	0
U. mirabilis	0	0	1	1	1	1	1	1	1	0	0	0	0
U. nazgul	1	0	1	1	1	1	1	1	1	0	0	0	0
U. niger	0	1	0	1	1	0	1	1	1	0	0	0	0
U. noyesi	0	0	1	1	1	?	?	1	1	0	1	1	1
U. pallidus	0	0	1	1	0	1	1	1	1	0	0	0	0
U. parvimalis	0	0	1	1	1	1	1	1	1	0	0	0	0
U. pintoi	0	0	1	1	1	1	1	1	1	0	0	0	0
U. placoides	1	0	1	1	1	1	1	1	1	0	0	0	0
U. principalis	0	0/1	1	1	1	0	1	1	1	0	0	0	0
U. rimatus	0	0	1	1	1	0	1	1	1	0	0	0	0
U. similis	0	1	0	1	1	0	1	1	1	0	0	0	0
U. simplipenis	0	1	0	1	1	0	1	1	1	0	0	0	0
U. spicifer	1	0	1	1	1	0	1	1	1	1	1	1	1
U. taniae	0	0/1	1	1	1	0	1	1	1	0	0	0	0
U. thylacinus	1	0	1	1	1	1	1	1	1	0	0	0	0
U. vectis	0	0	0	1	1	0	1	1	1	0	0	0	0

Table 1. *Ufens* morphological matrix (continued).

	Characters												
Taxon	14	15	16	17	18	19	20	21	22	23	24	25	26
Ceratogramma masneri	0	0	0	0	0	1	0	0	0	0	0	1	?
Mirufens sp.	0	0	0	0	0	0	0	0	0	0	0	0	?
Monorthochaeta nigra	0	1	0	0	0	1	?	0	1	0	0	1	0
Brachyufens osborni	0	0	0	0	0	0	0	0	0	0	0	0	?
New Genus Botswana	1	1	0	0	0	0	0	0	0	0	0	0	?
U. acacia	1	1	0	0	0	0	0	0	1	0	0	1	0
U. acuminatus	1	1	0	0	0	0	0	0	1	0	0	1	0
U. aperserratus	1	1	1	0	0	0	1	0	1	0	0	1	0
U. apollo	1	1	0	0	0	0	0	0	1	0	0	1	0
U. austini	1	1	0	0	0	0	0	0	1	1	0	0	0
U. australensis	1	1	0	0	0	0	0	0	1	0	0	1	0
U. bestiolis	1	1	0	0	0	0	0	0	1	0	0	0	0
U. cardalia	1	1	0	1	0	1	0	0	1	0	0	1	0
U. ceratus	1	1	0	0	0	0	0	0	1	0	1	1	0
U. cupuliformis	1	1	0	0	0	0	0	0	1	0	0	1	0
U. debachi	1	1	0	0	0	0	0	0	1	0	0	0	0
U. decipiens	1	1	1	1	0	1	0	0	1	0	0	0	0
U. dilativena	1	1	0	0	0	0	0	1	1	0	0	1	0
U. dolichopenis	1	1	0	0	0	0	0	0	1	0	0	1	0
U. elimaeae	1	1	0	0	0	0	0	0	1	0	0	1	1
U. flavipes	1	1	0	0	0	0	0	0	1	0	0	0	0
U. foersteri	1	1	0	0	0	0	0	0	1	0	0	1	0
U. forcipis	1	1	0	0	0	0	0	0	1	0	0	1	0
U. gloriosus	1	1	1	1	1	1	1	0	1	0	0	0	0
U. hercules	1	1	0	0	0	0	0	0	1	0	0	1	0
U. invaginatus	1	1	0	0	0	0	0	0	1	0	1	1	1
U. kender	1	1	0	0	0	0	0	0	1	0	0	1	0
U. khamai	1	1	0	0	0	0	0	0	1	0	0	1	0
U. kurrajong	1	1	0	0	0	0	0	0	1	0	0	1	0
U. lanna	1	1	0	0	0	0	0	0	1	0	0	1	0
U. messapus	1	1	0	0	0	0	0	0	1	0	0	0	0
U. mezentius	1	1	0	0	0	0	0	1	1	0	0	1	0
U. mirabilis	1	1	0	0	0	0	0	0	1	0	0	0	0
U. nazgul	1	1	0	0	0	0	0	1	1	1	0	0	0
U. niger	1	1	0	0	0	0	0	0	1	0	0	0	0
U. noyesi	1	1	1	1	0	0	0	0	1	0	0	1	0
U. pallidus	1	1	1	1	0	1	0	0	1	0	0	1	1
U. parvimalis	1	1	1	1	1	1	1	0	1	0	0	1	0
U. pintoi	1	1	0	0	0	1	0	1	1	0	0	1	0
U. placoides	1	1	1	1	1	1	1	0	1	0	0	1	0
U. principalis	1	1	0	0	0	0	0	0	1	0	0	0	0
U. rimatus	1	1	0	0	0	0	0	0	1	0	0	0	1
U. similis	1	1	0	0	0	0	0	0	1	0	0	0	0
U. simplipenis	1	1	0	0	0	0	0	0	1	0	0	1	0
U. spicifer	1	1	0	0	0	0	0	0	1	0	0	1	0
U. taniae	1	1	0	0	0	0	0	0	1	0	0	0/1	0
U. thylacinus	1	1	0	0	0	0	0	1	1	0	0	1	0
U. vectis	1	1	0	0	0	0	0	0	1	0	0	0	0

Table 1. *Ufens* morphological matrix (continued).

Taxon	27	28	29	30	31	32	33	34	35	36	37
Ceratogramma masneri	?	1	0	0	0	0	0	0	0	0	1
Mirufens sp.	?	0	?	?	0	0	0	0	0	0	1
Monorthochaeta nigra	0	1	0	0	0	0	?	0	?	?	?
Brachyufens osborni	?	0	?	?	0	0	0	0	0	0	1
New Genus Botswana	?	0	?	?	0	0	0	0	0	0	1
U. acacia	0	0	?	?	0	1	0	0	1	0	1
U. acuminatus	0	1	0	0	0	1	1	0	0	1	2
U. aperserratus	0	1	0	0	1	1	1	0	0	0	0
U. apollo	0	1	0	1	0	1	1	0	0	0	0
U. austini	1	1	0	0	1	1	1	0	0	0	0
U. australensis	1	1	0	0	1	1	1	0	0	1	0
U. bestiolis	0	1	0	0	1	1	0	0	0	0	1
U. cardalia	0	1	0	1	1	0	1	0	0	0	1
U. ceratus	0	0	?	?	1	1	0	0	0	0	2
U. cupuliformis	0	0	?	?	0	1	1	0	1	1	1
U. debachi	0	0	?	?	1	1	0	0	1	1	1
U. decipiens	0	1	0	0	1	0	0	0	0	0	0
U. dilativena	0	0	?	?	0	1	1	0	0	0	2
U. dolichopenis	0	0	?	?	0	1	0	0	0	0	2
U. elimaeae	0	0	?	?	0	1	1	0	1	1	2
U. flavipes	0	1	0	0	1	1	1	1	?	?	?
U. foersteri	0	0	?	?	0	1	1	0	0	0	0
U. forcipis	0	0	?	?	0	1	1	0	1	1	1
U. gloriosus	0	1	0	0	1	0	0	0	0	0	0
U. hercules	0	0	?	?	1	1	0	0	0	0	2
U. invaginatus	0	1	0	0	1	1	0	0	0	2	1
U. kender	0	1	1	0	1	1	0	0	0	0	0
U. khamai	0	1	0	1	0	1	1	0	1	1	1
U. kurrajong	0	1	0	0	1	0	1	0	0	2	1
U. lanna	0	0	?	?	0	1	0	0	0	2	1
U. messapus	1	0	?	?	1	1	0	1	?	?	?
U. mezentius	0	0	?	?	0	1	1	0	1	1	1
U. mirabilis	0	1	0	0	1	1	0	0	0	0	1
U. nazgul	0	1	1	0	1	1	1	0	1	2	2
U. niger	0	1	0	1	0	1	0	0	0	0	1
U. noyesi	1	1	0	0	1	1	1	0	0	0	1
U. pallidus	0	0	?	?	1	1	0	0	1	1	2
U. parvimalis	0	1	0	0	1	1	0	0	0	0	1
U. pintoi	1	1	0	0	1	0	0	0	0	0	0
U. placoides	0	1	1	0	1	0	1	0	0	0	0
U. principalis	0	1	0	1	0	1	0	0	0	0	1
U. rimatus	0	0	?	?	0	1	0	0	0	0	2
U. similis	0	1	0	1	0	1	0	0	0	1	2
U. simplipenis	0	0	?	?	0	1	0	0	0	0	2
U. spicifer	0	0	?	?	0	1	0	0	0	0	2
U. taniae	0	1	0	1	0	1	0	0	1	2	1
U. thylacinus	1	1	0	0	1	1	1	1	?	?	?
U. vectis	0	1	0	0	1	0	1	0	0	1	1

Results

All 37 morphological characters were parsimony-informative. Of the 1423 molecular characters, 1237 were constant, and, of the variable sites, 109 were parsimony uninformative and 77 parsimony informative.

Molecular Plus Morphological Analysis. In general, *Ufens* was recovered as monophyletic only in the analysis containing molecular data. All original MP analyses were poorly resolved (not illustrated individually), but SAW increased resolution. MP analysis utilizing both morphological and molecular data for those taxa with both data types yielded seven trees of length 373 in a single island (CI=0.702, RI=0.551). SAW analysis selected one of these seven trees (Fig. 1). This analysis of molecular and morphological data resulted in two groups, herein termed group A (*U. similis*, *U. principalis*, *U. dolichopenis*, *U. simplipenis*, *U. niger* and *U. taniae*) and group B (*U. bestiolis*, *U. kender*, *U. ceratus*, *U. foersteri*, *U. vectis* and *U. australensis*) (Fig. 1).

Morphological Analysis. No other analysis contained molecular data. Branch and bound analysis of only morphology of those taxa for which molecular and morphological data were present yielded 45 trees of length 52 in a single island (CI=0.577, RI=0.662). Ingroup monophyly was maintained in this analysis, but lacked resolution in the strict consensus (not illustrated). MP analysis of the morphology of all taxa yielded 78 trees of length 140 in a single island (CI=0.279, RI=0.620). Subsequent SAW yielded 144 trees of length 141 in a single island (Fig. 2), with *Ufens* paraphyletic. When ingroup taxa were constrained as monophyletic, MP analysis resulted in 36,146 trees of length 141 in 193 islands (CI=0.277, RI=0.617), without any ingroup resolution in the strict consensus. SAW yielded 7 trees of length 144 in a single island (Fig. 3). Backbone constraint of the morphological analysis (with the results of the strict consensus of the morphology + molecular data) resulted in 1,572 trees of length 161 in 45 islands (CI=0.242, RI=0.541). SAW yielded 48 trees of length 163 in a single island (Fig. 4A). When this backbone constraint was repeated without New Genus Botswana, 792 trees of length 157 in 39 islands were recovered (CI=0.248, RI=0.546). SAW yielded 2 trees of length 159 in a single island (Fig. 4A). The latter two analyses generated nearly identical phylogenetic hypotheses (Fig. 4A and 4B).

Discussion

These phylogenetic analyses should be considered preliminary. Not only is the morphological data set small relative to the number of species analyzed, but the level of homoplasy of the characters used is considerable. The heavy reliance on male genitalic features in this study may also be problematic. If genitalic differences in this genus are due to sexual selection as suspected (see discussion in Owen et al. 2007), the phylogenetic signal carried by this diversity is likely to be minimal.

Phylogenetic Analysis. Analysis of the *Ufens* species data yields highly labile results that are dramatically dependent upon analytical parameters. In fact, only analysis of the reduced morphological+molecular data set produced well-resolved hypotheses both with and without SAW (Fig. 1). The relationships recovered in this analysis were identical with those in the larger analysis of molecular data for the entire family (Owen et al. 2007), and *Ufens* monophyly is supported with 100% bootstrap support (Fig. 1). In fact, analysis of this reduced data set recovered groups A and B with nearly identical topology regardless of whether the morphological data were included (Fig. 1) or not (not illustrated). Clearly, the molecular characters possess clearer phylogenetic signal, and were the driving force in these results. Considering the comparative robustness of these results, it was believed highly justifiable to use this analysis as a backbone to help constrain the morphological data in a way consistent with the molecular results.

Of the two groups (A and B) generated by the analysis of molecular and morphological data (Fig. 1), only certain representatives of group A are regularly recovered as monophyletic in the morphology-only analyses (Figs. 2, 3). Unambiguous character state changes suggest the following morphological evidence for the recognition of group A: (1) relatively widely diverging fore wing setal tracks r-m and M (character 2); (2) male genitalia with a dorsal ridge (character 25); (3) male genitalia with base of ventral process greater than half the width of the capsule at base of process (character 30) (Fig. 1, 4A). Additionally, most of these taxa possess similar male genitalia (see below). Group B is somewhat less strongly corroborated, by the following unambiguous character state changes: (1) fore wing with a single setal track between setal tracks CU1 and CU2 (character 3) (Fig. 1); (2) male genitalia possessing a transverse hinge (character 31) (Fig. 1); (3) male genitalia with paramere base distinctly anterior to posterior edge of anterodorsal aperture (character 37) (Fig. 4A). Members of this group are known to possess rather distinctive male genitalia (see below). However, these character states are also shared by several species clearly not allied to either of these groups (Fig. 4A). Due to the absence of convincing resolution, no subgeneric categories are proposed.

The problem of anomalous results generated by limited morphological characters is well-documented (Scotland et al. 2003, Weins 2004), and the analysis of *Ufens* appears to be no exception. Perhaps these problems are not surprising, especially in light of the fact that no single morphological synapomorphy of *Ufens* species was found in the course of this study. Rather, as is characteristic of many genera of this family, *Ufens* is recognized by a suite of characters (see below). Nevertheless, the genus is easily recognized by the characters presented below, and all molecular analyses of the Trichogrammatidae point to monophyly (Owen et al. 2007), supporting its continued recognition. Although unique molecular synapomorphies of *Ufens* were not found, several unambiguous changes in 28S-D2 unite *Ufens* species (characters: 293 (unambiguous), 505 (region of slip strand compensation) and 817 (region of ambiguous alignment)) (Owen et al. 2007, pers. obs.).

Ufens monophyly was not supported in exclusively morphological analyses (Fig. 2). Its monophyly was disrupted by the intrusion of outgroup taxa, especially New

Genus Botswana (see below) unless constrained as monophyletic (Fig. 3), or incorporated into a backbone constraint without New Genus Botswana (Fig. 4). Nevertheless, these analyses point to several unambiguous morphological character state changes which may be important in most *Ufens* species' recognition: (1) fore wing with costal cell containing two distinct setal tracks (character 5); (2) male antenna with a four-segmented club (character 9); (3) male antenna with flagelliform setae arranged in a whorl (character 14); aedeagal apodemes absent (character 32) (Fig. 1, 4A). Some of these character states are shared by outgroup taxa, e.g. *Mirufens* and New Genus Botswana have males with four-segmented clubs. Additionally, a couple of these characters, namely the presence of two distinct setal tracks in the costal cell and of aedeagal apodemes, are known to vary within *Ufens*.

Unweighted parsimony analysis largely resulted in unresolved trees. SAW increased the level of resolution (Figs. 2-4), but in all except the reduced data set, at a cost of additional step changes. Therefore, the trees presented for the purely morphological data set are not the most parsimonious solutions. Particularly problematic in the unconstrained morphological analyses is that *Ufens* is never recovered as monophyletic (Fig. 1, 4A). In order to evaluate potential problems caused by this intrusion of outgroups, ingroup monophyly was constrained, requiring a single extra step in MP analysis (Fig. 3). Overall, New Genus Botswana was the most consistent ingroup intruder (Figs. 2, 4A). This taxon was included in analyses due to its morphological similarities to *Ufens*. However, it is clearly not an *Ufens* species due to its more primitive genitalia and long, sweeping stigmal vein. It was nested well-within *Ufens* due to multiple characters that it shares with *Ufens* species, such as the presence of flagelliform setae on the first funicular segment, and a hind wing width which increases distal to the hamuli (Fig. 4A). The possibility that this taxon is actually an aberrant *Ufens* cannot be completely excluded, though this is not considered likely (see below). Molecular data for this taxon would be especially useful to help evaluate its position in the Trichogrammatidae. When this taxon was eliminated in the backbone analyses, the outgroup was recovered as monophyletic without constraint (Fig. 4B). Most importantly, the phylogenetic hypotheses generated by these analyses, differing by only single outgroup taxon, are nearly identical (Fig. 4A and B), indicating that this taxon is not inducing radically different hypotheses when included.

Biogeography. The lack of conclusive evidence for any particular tree topology makes generalizations about biogeography tenuous, though some trends can be discerned. Most obviously, the taxa comprising group A are Holarctic; five of the six species in the group are Nearctic. Although it lacks molecular data, the Central American *U. apollo* has very similar male genitalia to certain members of this assemblage and probably belongs here as well. The only other New World species, *U. ceratus* and *U. debachi*, show no affinity with group A. Group B is comprised of primarily Australian species, with the Nearctic *U. ceratus* (Nearctic) nested within them (Fig. 1). The remaining species are widely scattered throughout the world

(Table 3), although most are from Australia. The phylogenetic hypotheses generated herein do not allow for a conclusive biogeographic hypothesis to account for their distribution.

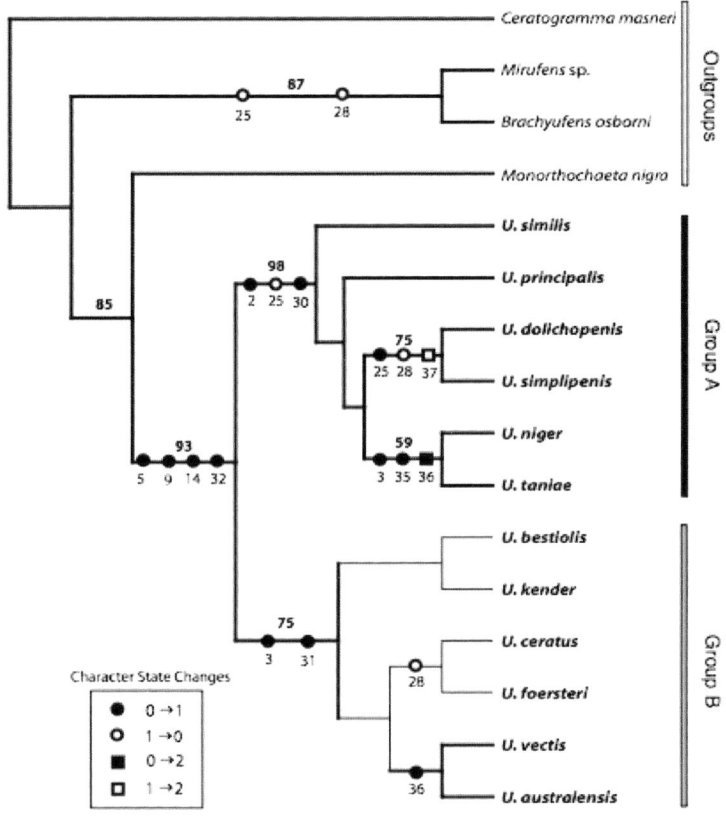

Figure 1. Successive approximations weighting cladogram of molecular and morphological data (TL=373, CI=0.702, RI=0.551). MP analysis yielded seven trees in a single island (TL=373). Only taxa with both types of data available were utilized in this analysis. Values above branches indicate bootstrap percentages greater than 50%. Unambiguous morphological character state changes are plotted below or to the left of branches. Branches in common with strict consensus of unweighted analysis are highlighted in bold. Ingroup monophyly was not constrained.

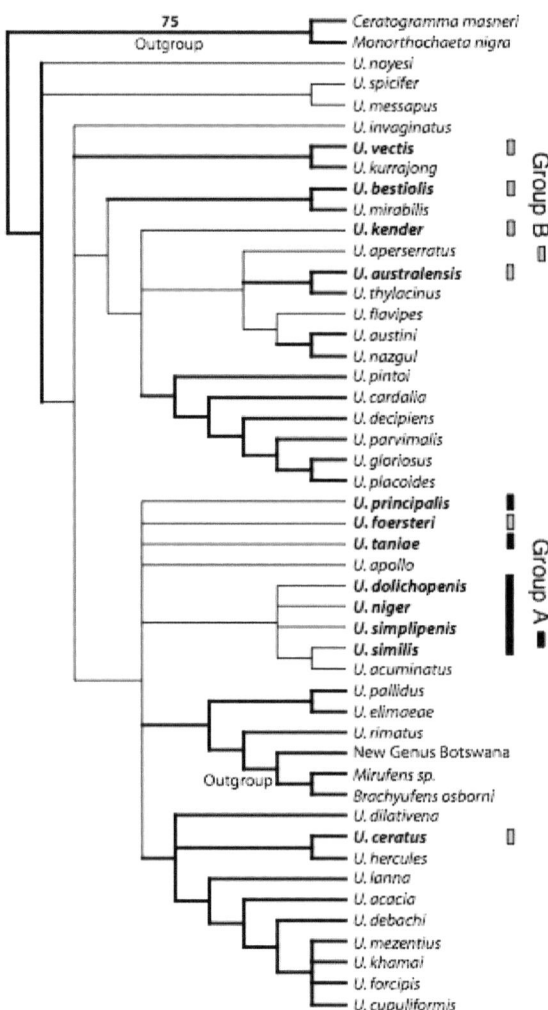

Figure 2. Successive approximations weighting cladogram of morphological data (TL=141, CI=0.279, RI=0.620). MP analysis yielded 78 trees in a single island (TL=140). Values above the branches indicate bootstrap percentages greater than 50%. Branches in common with strict consensus of unweighted analysis are highlighted in bold. Distributions of groups A and B from the analysis with molecular data (Fig. 1) are indicated. Note that *Ufens* is recovered as polyphyletic.

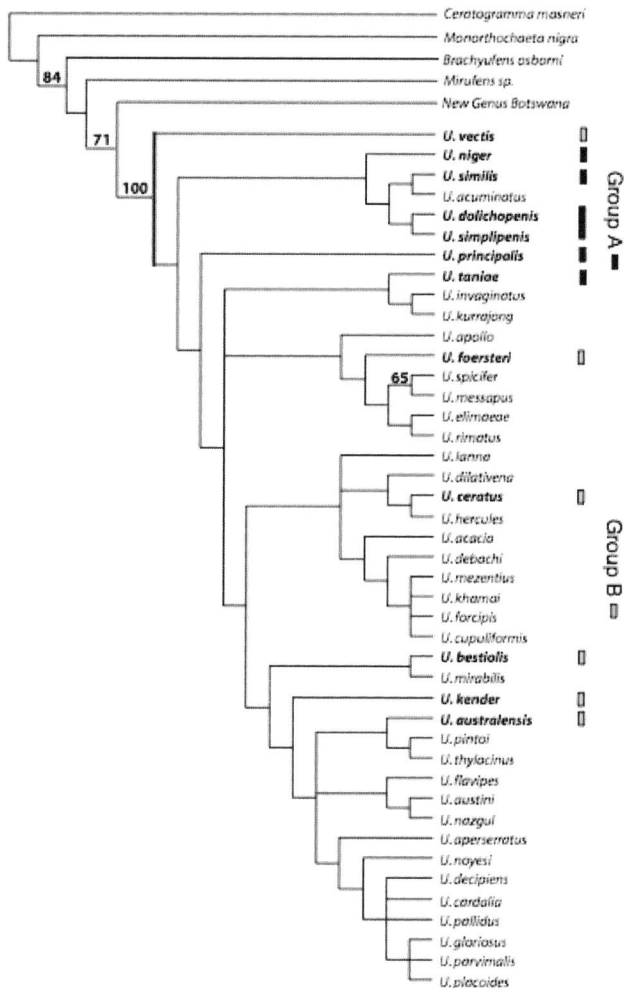

Figure 3. Successive approximations weighting cladogram of morphological data, with ingroup constrained monophyletic (TL=144, CI=0.277, RI=0.617). MP analysis yielded 36,146 trees in a 193 islands (TL=141). Values above the branches indicate bootstrap percentages greater than 50%. Branches in common with strict consensus of unweighted analysis are highlighted in bold. Distributions of groups A and B from the analysis with molecular data (Fig. 1) are indicated.

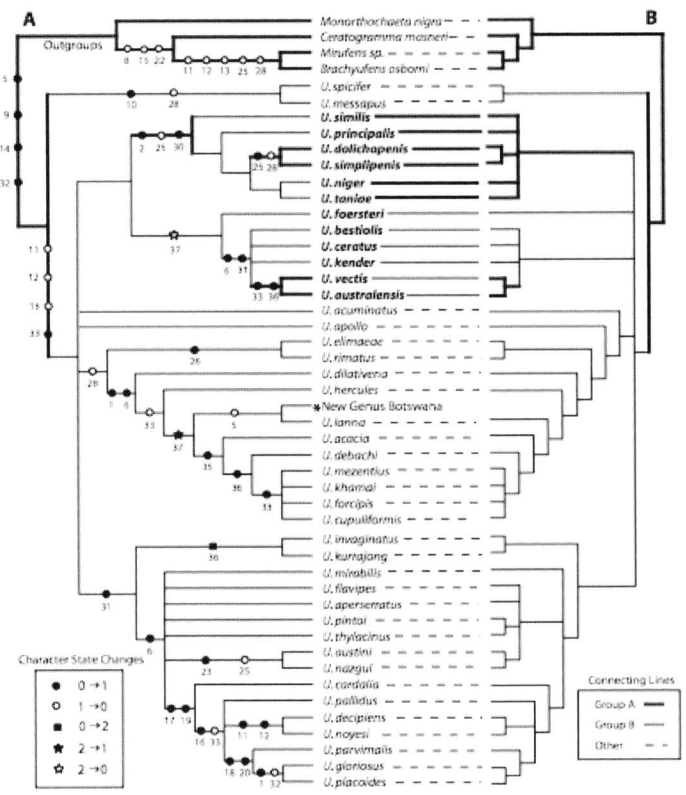

Figure 4. Successive approximations weighting cladogram of morphological data, using backbone constraint of the strict consensus of morphology + molecular data (cf. Fig. 1). A) SAW consensus of 48 trees (TL=163, CI=0.242, RI=0.541), with all taxa included. MP analysis yielded 1,572 trees in 45 islands (TL=161). B) SAW consensus of 2 trees (TL=159, CI=0.248, RI=0.546), with New Genus Botswana (indicated with a *) deleted. MP analysis yielded 792 trees in 39 islands (TL=157). Branches in common with strict consensus of respective unweighted analysis are highlighted in bold. Unambiguous morphological character state changes of internal branches are plotted below or to the left of branches. Distributions of groups A and B from the analysis with molecular data (Fig. 1) are indicated.

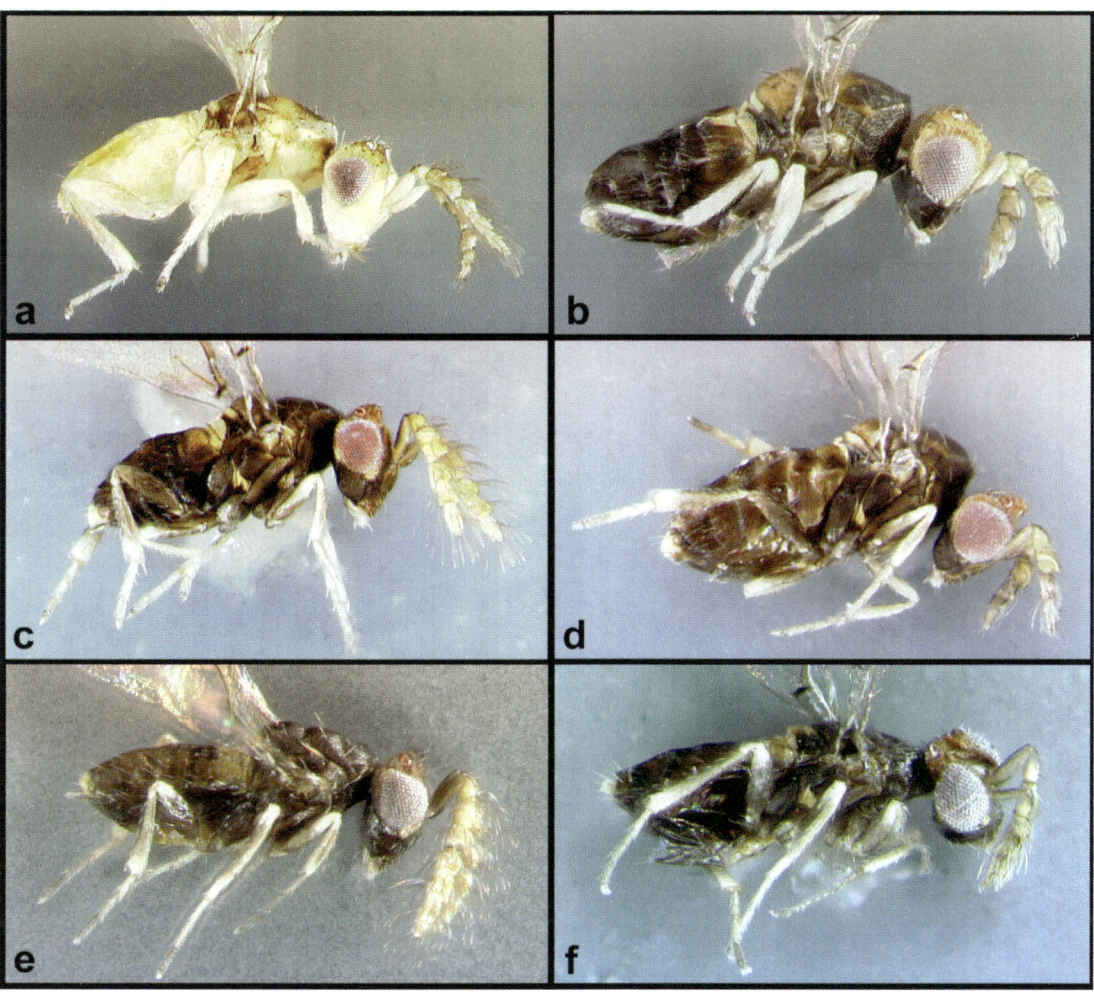

Figure 5. *Ufens* spp. habitus. (a) *U. ceratus,* ♂; (b) *U. ceratus,* ♀; (c) *U. principalis,* ♂; (d) *U. principalis,* ♀; (e) *U. simplipenis,* ♂; (f) *U. simplipenis,* ♀.

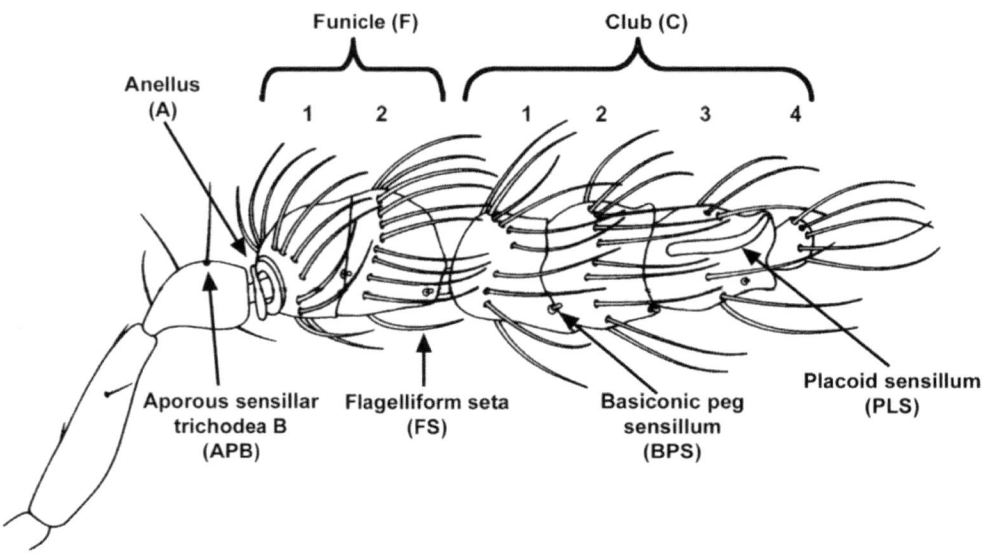

Figure 6. Generalized *Ufens* ♂ antenna (lateral), illustrating typical structures.

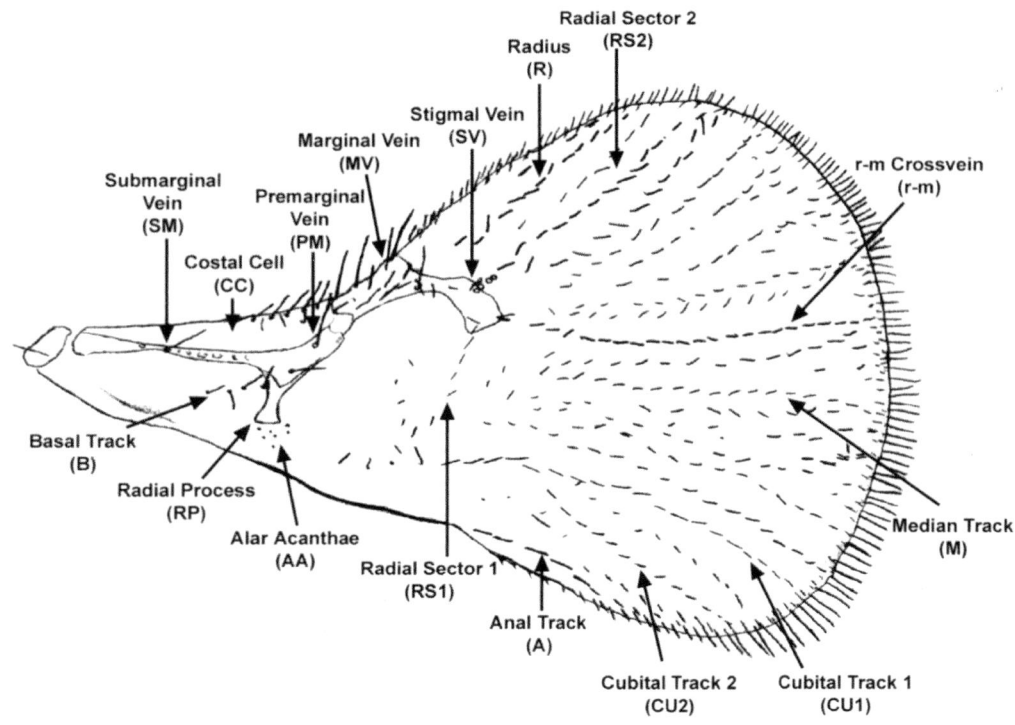

Figure 7. Generalized *Ufens* forewing (dorsal), illustrating typical structures and setal tracks.

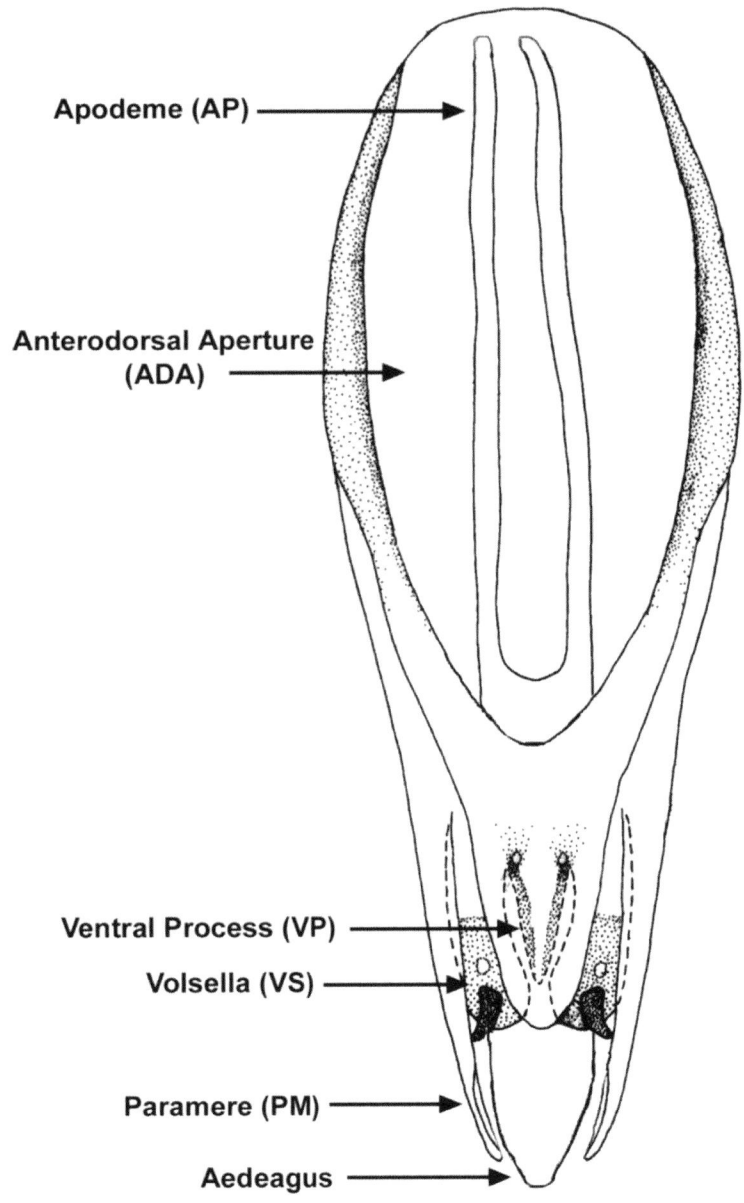

Figure 8. Generalized *Trichogramma* ♂ genitalia, illustrating typical placement and form of genitalic structures for most basal members of the Trichogrammatidae.

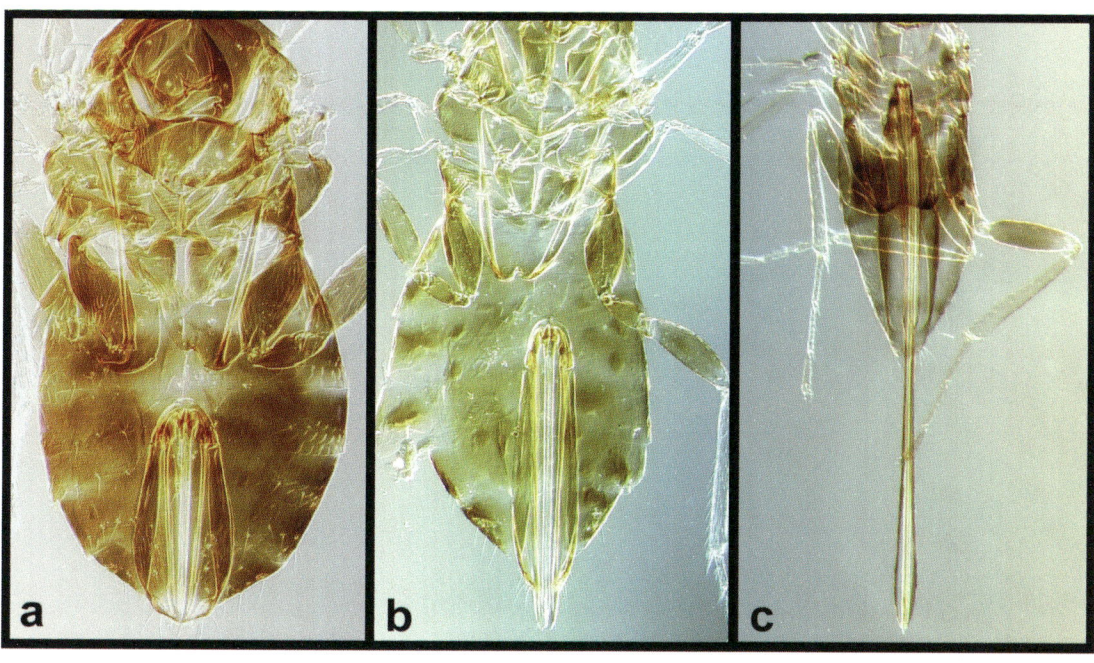

Figure 9. *Ufens* spp. ♀ ovipositors. (a) *U. simplipenis*; (b) *U. dolichopenis*; (c) *Ufens* sp. (Queensland, Australia). Note the varying lengths.

Ufens Revision

Ufens Girault

Ufens Girault 1911a: pp. 32-35. Type species: *Trichogramma nigrum* Ashmead, 1888: p. 107, by monotypy.

Ufensia Girault, 1913: p. 101. Type species: *Ufensia pretiosa* Girault, by monotypy. Girault, 1914: p. 118 (redescription). Doutt and Viggiani, 1968: pp. 576-578 (as synonym of *Ufens*). Viggiani 1972: pp. 159-161 (as valid genus). **Synonymy reinstated**.

Neocentrobia Blood, 1923: p. 254. Type species: *Neocentrobia hirticornis*, by monotypy. Blood and Kryger, 1928: p. 203 (redescription).

Stephanotheisa Soika, 1931: p.111. Type species *Stephanotheisa vitoldi*, by monotypy.

Grantanna Girault, 1939: p. 324. Replacement name for *Neocentrobia* Blood, 1923: p. 254; *nec* Girault, 1912, pp. 91-92.

Diagnosis. - Antennae with 2 anelli, 2 broadly joined funicular segments with PLS, 3 club segments in females, and 4 in males, including a comparatively small terminal segment. FS of male antenna arranged in whorls. Fore wing broad, often nearly truncate apically, with the MV and SV generally subequal in length. Major setal tracks discernible. Maxillary palp 1-segmented. Aedeagus fused to genital capsule.

The antennal formula of males (2 funicular and 4 club segments), 1-segmented maxillary palp, fore wing venation and male genitalia with fused aedeagus and capsule separate *Ufens* from all other trichogrammatid genera. Genera most likely confused with *Ufens* are *Mirufens*, New Genus Botswana, *Zagella*, *Chaetostricha* and *Japania*. In spite of some morphological similarities, those with molecular data do not appear to be closely related to *Ufens* based upon molecular evidence (Owen et al. 2007). *Mirufens* shares similar sexual dimorphism with *Ufens* in that its antennal club is also 4 segmented in males, but it is easily separated by its more sigmoid forewing venation, 2-segmented palp, and aedeagus that is separate from the genital capsule. New Genus Botswana in some ways represents an intermediate between *Mirufens* and *Ufens*. It shares a 4-segmented male club with both taxa, and possesses the long, sweeping stigmal vein and genitalia with aedeagus distinct from capsule characteristic of *Mirufens*. However, it shares with *Ufens* a 1-segmented maxillary palp and unridged pedicel. Placement of this new genus within the Trichogrammatidae is problematic, though its male genitalia clearly place it as a member of the Trichogrammatinae (cf. Owen et al. 2007). *Zagella*, *Chaetostricha*, and *Japania* lack sexually dimorphic antennae and the former two have male genitalia that are tube-like and almost always without parameres and volsellae. Although some species of *Ufens* possess highly simplified genitalia approaching a simple tube, either parameres or volsellae can be found in all known *Ufens* species except *U. flavipes*. Also, both *Zagella* and *Chaetostricha* are easily separated by funicular structure. In *Ufens*, the funicle segments are usually (but not always) subequal in length and F1 always bears at least one PLS. In

Zagella and *Chaetostricha*, F1 is considerably narrower and shorter than F2 and never bears PLS (Pinto 2006).

Etymology. - Not addressed by Girault (1911a), though the name may refer to an Italian (of Latium) warrior in Virgil's Aeneid (Book X and XII) named Ufens, who was killed in war by Gyas in 1176 B. C. It may also be in reference to the Ufens River, which ran below Setia, modern Seize, in ancient Latium, also mentioned in Virgil's Aeneid (Book VII, Ch. 38).

Distribution. - Worldwide (Table 3), though apparently uncommon in South America. The genus is most commonly collected in temperate and semi-arid regions. The following biogeographic regions, with number of species represented in parentheses, include: Afrotropical (5), Australian (23), Indomalaysian (8), Nearctic (6), Neotropical (3), Oceania (2) and Palearctic (6) (Table 3). The apparent level of sympatry within the genus can be fairly high, especially in Australia. Most species are known from a single or few biogeographic regions, though some species (e.g. *U. foersteri*, *U. vectis*) are known from three to four. No species is known to have a worldwide distribution (Table 3).

Biology. - Relatively little is known about the biology of *Ufens*, and the little that is known is from just a few species. *Ufens* is an arrhenotokous, gregarious, primary endoparasitoid in the eggs of other insects (Triapitsyn et al. 2002; Al-Wahaibi et al. 2005). Males of two species, *U. ceratus* and *U. principalis*, are known to emerge prior to females on a patch of hosts and compete for emerging females (Al-Wahaibi et al. 2005). The primary documented hosts for *Ufens* have been leafhoppers (Hemiptera: Cicadellidae) (Lin 1994), but they are also known from plant bugs (Hemiptera: Miridae) (Viggiani 1988) and katydids (Orthoptera: Tettigoniidae) (Timberlake 1927). Records by Peck (1963) from the eggs of moths (Lepidoptera: Pyralidae) need verification. This work recognizes host information for approximately eight species reared from at least fifteen hosts associated with a wide variety of host plants.

Description. - Body compact, 0.5 - 0.9 mm. long; BL/HTL = 2.6-5.2. Color varying from brown to yellow; eye red; gena usually darker than rest of head; antenna often lighter in color than rest of body; legs usually darker towards base, with tibiae and tarsi often very light, and each terminal tarsal segment often darker than others (Figure 5).

Head. Width subequal to mesosoma width; vertex wrinkled, not as well-sclerotized as gena and face. Postgena with narrow groove dorsally. Malar sulcus present. Mandible with approximately three acute, heavily sclerotized teeth posteriorly, and subcrenate anteriorly. Clypeus subquadrate. Maxillary palp 1-segmented. Antenna arising at mid to lower eye level; toruli separated by distance subequal to or slightly greater than their width; sexually dimorphic; both sexes with 2 anelli; 2 broadly joined funicular segments, with maximum length of F2 subequal to 1.8x maximum length of F1; funicular segments at least as wide as club segments.

Male antenna (Figure 6): 4 club segments, C4 distinctly smaller than all other segments; C/F = 1.5-3.2; F2/F1 = 0.6-1.5; FS arranged in whorls around each flagellar segment and the only sensillar type on C4, generally abundant on each of F1-C4, though often fewer on F1 and F2; US uncommon, but occasionally present on F1-C1; 3-7 APB on pedicel, usually absent on F and C; generally 1 PLS on F1-C2 and 2 on C3, but with up to 9 PLS on each of these segments; BPS associated with apex of each segment of F and C except C4, usually more numerous on F1, F2, and C1, usually only one BPS on C2 and C3.

Female antenna: 3 club segments; C/F = 1.6-2.7; F2/F1 = 0.7-2.9, but F2 length usually subequal to or slightly greater than in F1; APB located dorsally when present, 1 APB on F1 and F2, 0 APB on C1 and C2, 0-1 APB on C3; PLS number variable but generally with similar count on each segment F2-C2 and a greater number on C3, 1-2 on F1 but sometimes up to 4, 1-10 on F2, 1-5 on C1 and C2, 3-6 on C3; BPS associated with apex of each segment of F and C, 1-8 on each segment F1-C1, with a generally similar count on each of these segments within an individual, 1 BPS on C2 and C3, primarily located ventrally; FS absent on F1 and F2, 4-11 on C1, 5-17 on C2, 2-8 on C3, often most abundant on C2 and usually located dorsoapically; 0-1 UPP on C3; US abundant on all C and F segments except C2, 4-17 on each of F1-C1, 0 on C2, 0-9 (rarely > 5) on C3.

Mesosoma. Pronotum divided medially. Mesopleural suture present. Midlobe of mesoscutum and scutellum each with two pairs of setae; anterior pair on scutellum sometime shorter. Legs with short, straight protibial spur, anterior of metatarsus with basitarsal setal comb. Fore wing (Figure 7) broad, often nearly truncate apically and length approximately 3x HTL; venation somewhat elongate, reaching approximately half FWL; setal density, as inferred by counting number of setae between RS_2 and r-m, usually 50-200; SV generally with a distinct constriction at base; less-sclerotized between PM and MV; RP distinct, with varying levels of sclerotization, generally with blunt termination; setae in CC generally arranged in 2 distinct tracks; basal vein track generally present; vein tracks R, RS2, r-m, M, CU1, CU2, RS1 all distinguishable, with varying levels of setation between them; two campaniform sensilla located near apex of PM, sometimes adjacent but usually separated by at least their width; AA present or absent posterior to termination of radial process; fringe setae short, generally longest at posterolateral margin; FWL/HTL = 2.6-3.2; FWL/FWW = 1.3-3.9; FWFS/FWW = .02-.18; Max r-m to M/Min r-m to M = 0.3-6.5; MV/PM = 0.6-1.7; SV/MV = 0.6-2; MV length/MV width = 1.5-5.3. HW generally with 3 distinct setal tracks, widest at hamuli to mid-disk; HWL/HWW = 5.6-11.2, HWFS/HWW = 0.6-1.5. Mesophragma notched posteriorly. Mesofurca with lateral arms tapering, sigmoid distally.

Metasoma. Male genitalia variable, aedeagus indistinct from genital capsule; ADA present; AP/GL = 0.2-0.6, though generally not present; PAR, VS, and VP generally present; PAR/GL = 0.1-0.8; VS/GL = 0.1-0.6; VP/GL = 0.1-0.9; DR, transverse hinge present or absent; GL/GW = 1.1-5.8; GL/HTL = 0.6-1.6; AI/GL = 0-0.3; ADA/GL = 0.2-0.7; PM/GL = 0.1-0.7. Ovipositor generally not extending appreciably beyond apex of metasoma, OL/HTL = 0.8-3.9.

Comments. - All generic names listed above under *Ufens* Girault were considered junior synonyms by Doutt and Viggiani (1968). Only *Ufensia* was more recently considered a valid genus. A complete list of species recognized herein is found in Table 2.

Ufensia was originally described by Girault (1913) and distinguished from *Ufens* by minimal characteristics such as ovipositor length. It had been considered a junior synonym of *Ufens* (Doutt and Viggiani 1968) until it was resurrected by Viggiani (1972), based upon the reduced male genitalia with a relatively small anterodorsal aperture. It has become clear that Viggiani did not have the taxa necessary to appreciate the true diversity of *Ufens* genitalia, which have variation matching or exceeding that found throughout the rest of the family (cf. Figs. 10 – 52). The complex is otherwise extremely conservative anatomically and no structures other than the genitalia suggest a partitioning of *Ufens*. The taxa placed into *Ufens* and *Ufensia* simply exhibit two of several "genitalic themes" which occur in this group. They are not extraordinary when enough species are examined, as intermediates and other radically different themes also occur. This confusion highlights the danger both of incomplete collections and the use of single-character systems to classify taxa in this family. In fact, the previous higher-level classification placed sections of this single genus into both subfamilies: *Ufens* in the Trichogrammatinae and *Ufensia* in Oligositinae (Viggiani 1971).

Seven species are currently incorrectly assigned in *Ufens*. These species were originally described in *Mirufens* and are herein reassigned to that genus. Included are *Mirufens albiscutellum* Khan and Shafee, *M. brevifuniculata* Khan and Shafee, *M. longiclavata* Khan and Shafee, *M. magniclavata* Khan and Shafee, *M. afrangiata* Viggiani and Hayat, *M. longifuniculata* Viggiani and Hayat, and *M. mangiferae* Viggiani and Hayat (**revised combinations**). All were treated as new combinations in *Ufens* by Yousuf and Shafee (1987). However, in their review of *Mirufens*, Neto and Pintureau (1997) do not acknowledge these transfers. I have examined the types of all seven species and can confirm that they are indeed *Mirufens*; they are not further addressed in this work. A couple of other species, *Burksiella benefica* and *B. spirita* (as *Zagella*), have recently been transferred out of *Ufens* by Pinto (2006) and Triapitsyn (2003), respectively.

Although dissections were not performed, the mesofurca of *Ufens* is visible in high-quality slide mounted material. Heraty et al. (1997) described the mesofurca of trichogrammatids as having the mesofurcal bridge completely lost, and the lateral arms tapering, straight and projecting forward. The lateral arms of *Ufens* are more similar to the state Heraty et al. (1997) considered autapomorphic for *Xiphogramma*, in that their apex is curved in a sigmoid fashion. The shape of the mesofurca found in *Ufens* and *Xiphogramma* is very similar to the shape found in *Cales* (*incertae sedis*) and *Eretmocerus* (Aphelinidae) (Heraty et al. 1997).

Therefore, further information is needed over a broader range of trichogrammatid genera to determine if this structure provides phylogenetic signal or if it can be useful for identification.

Table 2. *Ufens* species list.

Ufens acacia Owen, **new species**
Ufens acuminatus Lin, 1993
Ufens aperserratus Owen, **new species**
Ufens apollo Owen, **new species**
Ufens austini Owen, **new species**
Ufens australensis Owen, **new species**
Ufens bestiolis Owen, **new species**
Ufens cardalia Owen, **new species**
Ufens ceratus Owen, 2005
Ufens cupuliformis Lin, 1993
Ufens debachi Owen, **new species**
Ufens decipiens Owen, **new species**
Ufens dilativena Nowicki, 1940, **revised combination**
Ufens dolichopenis Owen, **new species**
Ufens elimaeae Timberlake, 1927
Ufens flavipes Girault, 1912
Ufens foersteri (Kryger, 1918)
        Syn.: *Ufens hirticornis* (Blood), 1923
        Syn.: *Ufens foersteri irregularis* Nowicki, 1935
        Syn.: *Ufens foersteri meridionalis* Nowicki, 1935
        Syn.: *Ufensia africana* Viggiani, 1972
        Syn.: *Ufensia minuta* Viggiani, 1988
Ufens forcipis Owen, **new species**
Ufens gloriosus Owen, **new species**
Ufens hercules Girault, 1912
Ufens invaginatus Owen, **new species**
Ufens kender Owen, **new species**
Ufens khamai Owen, **new species**
Ufens kurrajong Owen, **new species**
Ufens lanna Owen, **new species**
Ufens messapus Owen, **new species**
Ufens mezentius Owen, **new species**
Ufens mirabilis Owen, **new species**
Ufens nazgul Owen, **new species**
Ufens niger (Ashmead, 1888)
Ufens noyesi Owen, **new species**
Ufens pallidus Owen, **new species**
Ufens parvimalis Owen, **new species**
Ufens pintoi Owen, **new species**
Ufens placoides Owen, **new species**

Table 2. *Ufens* species list (continued)

Ufens principalis Owen, 2005

Ufens rimatus Lin, 1993

Ufens similis (Kryger, 1932)

       Syn.: *Ufens similis megaloptila* Nowicki, 1940

       Syn: *Ufens anomalus* Lin, 1994, **new synonomy**

Ufens simplipenis Owen, **new species**

Ufens spicifer Owen, **new species**

Ufens taniae Owen, **new species**

Ufens thylacinus Owen, **new species**

Ufens vectis Owen, **new species**

Nomina dubia

Ufens albitibiae Girault, 1915

Ufens alami Yousuf and Shafee, 1987

Ufens angustipennis Yousuf and Shafee, 1987

Ufens binotatus Girault, 1915

Ufens breviclavata Yousuf and Shafee, 1991

Ufens gurgaonensis Yousuf and Shafee, 1987

Ufens jaipurensis Yousuf and Shafee, 1987

Ufens latipennis Yousuf and Shafee, 1987

Ufens luna Girault, 1916

Ufens piceipes Girault, 1912

Ufens pretiosus (Girault, 1913)

Ufens quadrifasciatus Girault, 1915

Ufens singularis Yousuf and Shafee, 1987

Table 3. Known geographic distribution of recognized *Ufens* species (darkened cells). Biogeographic realms according to Olson et al. 2001 (available at http://www.worldwildlife.org/science/ecoregions/WWFBinaryitcm6498.pdf).

SPECIES	Afrotropical	Australasian	Indomalaysian	Nearctic	Neotropical	Oceania	Palearctic
U. acacia	■						
U. acuminatus			■				
U. aperserratus		■					
U. apollo					■		
U. austini		■					
U. australensis		■					
U. bestiolis		■					
U. cardalia		■					
U. ceratus				■			
U. cupuliformis			■				
U. debachi				■			
U. decipiens		■					
U. dilativena	■						■
U. dolichopenis				■			
U. elimaeae						■	
U. flavipes		■					
U. foersteri	■	■	■				■
U. forcipis							■
U. gloriosus		■					
U. hercules		■					
U. invaginatus		■					
U. kender		■					
U. khamai	■		■				
U. kurrajong		■					
U. lanna			■				
U. messapus		■					
U. mezentuis			■				■
U. mirablis		■					
U. nazgul		■					
U. niger				■			
U. noyesi		■					
U. pallidus							■
U. parvimalis		■					
U. pintoi		■					
U. placoides		■					
U. principalis				■			
U. rimatus			■				
U. similis	■						■
U. simplipenis				■			
U. spicifer		■					
U. taniae					■		
U. thylacinus		■					
U. vectis		■		■		■	

Key to *Ufens* Species of the World
(males)

1.   Aedeagal apodemes (AP) present (Figs. 21d, 33d) 2
1'.   AP absent (Figs. 31e, 48i).. 8

2 (1).   Ventral process (VP) present, width at base greater than half the genital
        capsule width; parameres (PAR) short and thin; anterior invagination (AI)
        prominent (Fig. 17d) (Australia) ....*U. cardalia*
2'.   VP absent (Figs. 21d, 28d) or, if present, width at base of VP less than half
        the genital capsule width (Figs. 17d, 52e, f); PAR and AI variable 3

3 (2').   First club segment (C1) with 1 placoid sensillum (PLS) or lacking PLS
        entirely (Figs. 21a, 52c) 4
3'.   C1 with more than 1 PLS (Figs. 28a, 44a) 7

4 (3).   Second club segment (C2) with 1 placoid sensillum (PLS) or lacking PLS
        entirely (Figs. 33a, 52c) 5
4'.   C2 with more than 1 PLS (Fig. 28a) 6

5 (4).   Fore wing (FW) setation sparse. Parameres (PAR) widest near middle;
        ventral process (VP) long (Fig. 33b, d) (Australia) .... *U. kurrajong*
5'.   FW setation dense. PAR widest near base; VP short (Fig. 52a, e-f)
        (Australia, New Guinea, New Zealand, Indonesia) .... *U. vectis*

6 (4').   Volsellae (VS) not reaching genitalic apex, connected basally, and curving
        towards midline (Fig. 21d) (Australia) .... *U. decipiens*
6'.   VS reaching genitalic apex, not connected basally, and straight (Fig. 43d)
        (Australia) .... *U. pintoi*

7 (3').   First funicle segment (F1) without flagelliform seta (FS). Ventral process
        (VP) entire; parameres (PAR) short, their length much less than 0.5x entire
        genitalia length (GL); volsellae (VS) short, connected basally, and curving
        towards midline (Fig. 28a, d) (Australia) .... *U. gloriosus*
7'.   F1 with FS. VP asymmetrically bifurcating in apical third; PAR long, their
        length about 0.5x GL; VS absent (Fig. 44d) (Australia) ...... *U. placoides*

8 (1').   Lateral margins of anterodorsal aperture (ADA) abruptly constricted at
        posterior half (Figs. 40c, 51d) 9
8'.   Lateral margins of ADA relatively uniformly tapering along entire length
        (Figs. 39f-g, 31d-e) 13

9 (8).   Ventral process (VP) absent (Fig. 35d) (Australia) .... *U. messapus*
9'.   VP present (Figs. 25f, 45h) 10

10 (9'). Parameres (PAR) absent; small appendages at apex of genitalia present
(Fig. 51d) (Australia) .... *U. thylacinus*
10'. PAR present; small appendages at apex of genitalia absent (Figs. 14d, 15d)
.... 11

11 (10'). Bulbous area posterior of transverse hinge present, dorsal ridge (DR)
present (Fig. 14d) (Australia) .... *U. austini*
11'. Bulbous area posterior of transverse hinge absent, DR absent (Figs. 15d,
40c) 12

12 (11'). Funicle without aporous sensillar trichodea B (APB) and with unsocketed
seta (US); first funicle segment (F1) without flagelliform seta (FS).
Parameres (PAR) subequal in width along entire length (excluding
terminal spine) (Fig. 40a, c) (Australia) .... *U. noyesi*
12'. Funicle with APB and without US; F1 with FS. PAR widest at base (Fig.
15d) (Australia) .... *U. australensis*

13 (8'). Hind wing (HW) width precipitously narrowing immediately beyond
hamulus (Figs. 11c, 39d) 14
13' HW width maintained or increasing immediately beyond hamulus (Figs.
20d, 31c) 26

14 (13). Anterior invagination (AI) pronounced, its length greater than 0.2x
genitalia length (GL) (Fig. 30d) (Australia) .... *U. invaginatus*
14'. AI shallow, its length less than 0.2x GL (Figs. 20h, 45f-h) 15

15 (14'). Fourth club segment (C4) minute, not extending beyond placoid sensilla
(PLS) on third segment (C3); club compact, suboval and without obvious
constrictions between segments. Volsellae (VS) long and spine-like (Fig.
49a, d) (Australia) .... *U. spicifer*
15'. C4 normal, clearly extending beyond PLS on C3; club variable, generally
with obvious constrictions between segments and somewhat elongate. VS,
when present, variable (Figs. 11a,d, 45a,h) 16

16 (15') Paramere (PAR) without terminal spine (Figs. 50e, 24f) 17
16'. PAR with terminal spine (Figs. 45h-i, 31e-f) 18

17 (16). Anterodorsal aperture (ADA) with maximum width subequal to capsule
maximum width; parameres (PAR) apically spatulate (Fig. 50e) (Central
and South America) .... *U. taniae*
17'. ADA width distinctly narrower than capsule width; PAR apically tapering
(Fig. 24e-f) (Hawaiian Islands) .... *U. elimaeae*

18 (16'). Ventral process (VP) absent (Figs. 46d, 48j) 19
18'. VP present (Figs. 39h, 45h) 22

19 (18). Anterodorsal aperture (ADA) with maximum width distinctly narrower than capsule maximum width (Fig. 46d) (Asia) .... *U. rimatus*

19'   ADA width subequal to capsule width (Figs. 26f-g, 48h-i) 20

20 (19'). Paramere (PAR) long, greater than 0.5x genitalia length (GL), its base anterior of posterior-most edge of anterodorsal aperture (ADA); ADA short, its length approximately 0.3x GL; anterior invagination (AI) notch-like (Fig. 26f-i) (Afrotropics, Australia, Indomalaysia, Palearctic) .... *U. foersteri*

20'.   PAR short and difficult to discern, only ca. 0.1x GL, their base posterior of posterior-most edge of ADA; ADA longer, its length ca. 0.5x GL; AI indistinct, not notch-like (Figs. 23d, 48h-k) 21

21 (20'). Genitalia length (GL) less than (0.8-0.9x) hind tibial length (HTL) (Fig. 48h-k) (central and western U.S., México) .... *U. simplipenis*

21'.   GL greater than (1.2-1.5x) HTL (Fig. 23d) (western U.S.) .... *U. dolichopenis*

22 (18'). Ventral process (VP) base less than 0.5x width of capsule at base of process, volsellae (VS) absent (Fig. 11d) (China, Thailand) .... *U. acuminatus*

22'.   VP base greater than 0.5x width of capsule at base of process, VS variable but usually present (Figs. 39f-h, 45f-i) 23

23 (22'). Forewing (FW) setation sparse. Dorsal ridge (DR) absent. (Fig. 13a, e) (México and Central America) .... *U. apollo*

23'.   FW setation dense. DR present (Figs. 39a,f, 47a,d) 24

24 (23'). Ventral process (VP) laterally emarginate on one side (Fig. 47d) (Africa, Palearctic) .... *U. similis*

24'.   VP laterally entire (Figs. 39h, 45h) 25

25 (24'). Midlobe of mesoscutum with sculpturing cellulate. Genitalia with anterior margin usually rounded and not emarginated, capsule curving slightly, subsinuate (Fig. 39e, f-h) (Nearctic, Carribean) .... *U. niger*

25'.   Midlobe of mesoscutum with sculpturing longitudinally striate. Genitalia with anterior margin usually more transverse and often slightly but noticeably emarginate, capsule relatively straight entire length, not subsinuate (Fig. 45e, f-i) (western United States) .... *U. principalis*

26 (13'). Ventral process (VP) present (Figs. 25f, 31f) 27

26'.   VP absent 34

27 (26). Paramere (PAR) and volsella (VS) absent (Fig. 25d-f) (Australia)
.... *U. flavipes*
27'.   PAR present, VS present or absent (e.g. Figs. 31d-g, 32d-f) 28

28 (27'). Paramere (PAR) without terminal spine (Figs. 32d-f, 38d) 29
28'.   PAR with terminal spine (e.g. Figs. 12e, 31d-f) 30

29 (28). Parameres (PAR) not broadly flattened dorsoventrally, diverging apically
          from midline, with minute perforations apically; ventral process (VP) short
          and entire; transverse hinge absent (Fig. 32d-g) (Botswana, India)
          .... *U. khamai*
29'.   PAR flattened dorsoventrally, not diverging apically, curving slightly
          towards midline and without perforations; VP asymmetrically bifurcating
          (difficult to see), transverse hinge present (Fig. 38d) (Australia)
          .... *U. nazgul*

30 (28'). Volsellae (VS) absent; capsule floor fenestrate in apical portion of basal
          third, split longitudinally in apical 2/3, median sides of split serrate (Fig.
          12d) (Australia) .... *U. aperserratus*
30'.   VS present; capsule floor normal, not fenestrate or longitudinally split (e.g.
          Figs. 31d-g, 42d) 31

31 (30'). Ventral process (VP) elongate, its base attaining basal margin of capsule
          and with obvious opening on capsule floor; anterodorsal aperture (ADA)
          'heart' shaped and extending ca. 0.5x genitalia length (GL) (Fig. 31d-g)
          (Australia) .... *U. kender*
31'.   VP shorter, its base not extending beyond posterior half of capsule and
          without obvious opening on capsule floor; ADA variable but not as above
          32

32 (31'). Both funicle segments with at least 2 placoid sensilla (PLS) (Fig. 42a)
          (Australia) .... *U. parvimalis*
32'.   Both funicle segments with 1 PLS (Figs. 37a, 16a) 33

33 (32'). Anterior invagination (AI) indiscernible; parameres (PAR) with terminal
          spine thick and heavily sclerotized (Australia) (Fig. 37d) .... *U. mirabilis*
33'.   AI prominent, its depth greater than 0.05x genitalia length; PAR with
          terminal spine thin, lightly sclerotized (Australia) (Fig. 16d) . *U. bestiolis*

34 (26'). Paramere (PAR) with terminal spine (Figs. 18g-j, 34d)   35
34'.   PAR without terminal spine (Figs. 10e, 41d) 38

35 (34). Paramere (PAR) broad, widening near middle (immediately basal of apical
          spine), its base even with posterior edge of anterodorsal aperture (ADA)
          (Fig. 34d) (Thailand) .... *U. lanna*

35'.   PAR much narrower, equally wide entire length, its base apical of posterior edge of ADA (e.g. Figs. 18g-j, 29d) 36

36 (35'). Genitalia with minute paired spines at apex (Fig. 29d) (Australia) .... *U. hercules*

36'.   Genitalia without paired spines at apex (Figs. 18g-j, 22d) 37

37 (37'). Head with forward-projecting stout setae on the lateral margins of clypeus and two adjacent pairs on genae. Anterior invagination (AI) pronounced, its length 0.2-0.3x genitalia length (GL); volsellae (VS) present (Fig. 18f, g-i) (United States and México) .... *U. ceratus*

37'.   Head without stout setae. AI shallow to nonexistent; VS absent (Fig. 22d) (Palearctic, Africa) .... *U. dilativena*

38 (34'). Forewing (FW) venation very pale; stigmal vein (SV) only slightly constricted at base; costal cell with very few setae, arranged in, at most, a single row (Fig. 41b) (Turkmenistan) .... *U. pallidus*

38'.   FW venation well pigmented; SV constricted at its base; costal cell with two distinct rows of setae (e.g. Figs. 10a, 20c) 39

39 (38'). Transverse hinge present; parameres (PAR) with apices dark and heavily sclerotized (Fig. 20g-h) (southwestern United States, Baja California) .... *U. debachi*

39'.   Transverse hinge absent; PAR without differentially sclerotized apices 40

40 (39'). Parameres (PAR) uniformly wide, flattened, spatuliform, sigmoid and curving towards midline; volsellae (VS) present (Fig. 10e) (Botswana) .... *U. acacia*

40'.   PAR variable but not as above; VS absent (Figs. 19b, 36d) 41

41 (40'). Parameres (PAR) diverging from midline most of their length and laterally emarginate near apex (Fig. 19b) (China) .... *U. cupuliformis*

41'.   PAR straight or only diverging from midline at very apex, not laterally emarginate (Figs. 27d, 36d) 42

42 (41'). Parameres (PAR) straight most of their entire length, only slightly diverging from midline apically; anterodorsal aperture (ADA) twice as wide as long (Fig. 27d) (Oman) .... *U. forcipis*

42'.   PAR more strongly diverging from midline apically; ADA less than twice as wide as long (Fig. 36d) (Israel, Sri Lanka, South Africa)   ...... *U. mezentius*

Ufens Species Descriptions
[♦ indicates type material examined]

Ufens acacia Owen, new species
(Fig. 10)

Diagnosis. - Fore wing sparsely setose with narrowly diverging setal tracks r-m to M and a single setal track between CU1 and CU2. Hind wing width not decreasing immediately apical of hamuli. Mesoscutal sculpturing longitudinally striate. Genitalia with parameres wide, flattened, spatuliform, lacking a terminal spine, their base even with posterior edge of anterodorsal aperture, somewhat sigmoid, with their apical half curving towards midline; volsellae base stout, arising medially and ventrally, apex projecting dorsally, thin and curving.

This species is unlikely to be confused with any other due to the unique shape of the volsellae and parameres. The volsellae in particular are quite distinctive as their base is medial and ventral, and apically they project dorsally. The shape of the ventral portion of the capsule must somehow accommodate this projection, but is indiscernible in the material available. The curvature of the volsellae prohibits their accurate measurement, though the distance in a straight line from base to apex is subequal to the length of the parameres. No other species are known to have volsellae which project dorsally through the capsule. In fact, the only other species known to have an appendage apparently projecting dorsally through the capsule is *U. aperserratus*. In this case, however, it is the ventral process with its base ventral and apex dorsal. However, the two species have little else in common, as *U. acacia* does not have a ventral process and is further distinguished by its broad spatuliform parameres without a terminal spine.

Types. - ♦Holotype ♂. **BOTSWANA: Serowe**: Farmer's Brigade, x.1987, MT, P. Forchhammer (CNC). Paratypes 2♂, same data except one vii.1987, the other ix.1987 (UCRC).

Etymology. - Named after *Acacia* spp. (Fabaceae), among the most common trees in Botswana.

Distribution. - Botswana.

Biology. - Unknown.

Description (N=2). - BL 0.54 (0.53-0.55) mm. BL/HTL = 3.3 (3.1-3.5). Mesoscutal sculpturing longitudinally striate with interstitial sculpturing primarily transverse. Forewing sparsely setose; AA absent; single setal track between CU1 and CU2; FWL/HTL = 3.0 (2.8-3.1); FWL/FWW = 1.5; FWFS/FWW = 0.06; Max r-m to M/Min r-m to M = 1.7 (1.6-1.9); MV/PM = 1.1 (1.0-1.1); SV/MV = 1.0 (0.9-1.1); MV length/MV width = 2.5 (2.4-2.9). Hind wing width not decreasing immediately apical of hamuli; HWL/HWW = 7.8 (7.6-8.2); HWFS/HWW = 0.8.

Male

Antenna: C/F = 2.4 (2.3-2.5); F2/F1 = 1.5 (1.2-1.7); APB absent on funicle; 1 PLS on each of F1-C2, 2 PLS on C3; 3-4 BPS on each of F1-C1, 1 BPS on C2 and C3; 7-8 FS on F1, 9-11 FS on F2, 8-9 FS on C1, 8-10 FS on C2, 7-9 FS on C3, 5-6 FS on C4; US absent on each of F1-C3.

Genitalia: Capsule subtriangular and broad; GL/GW = 1.2 (1.1-1.3); GL/HTL = 0.5 (0.5-0.6); ADA/GL = 0.4 (0.4-0.5); AI pronounced, AI/GL = 0.1 (0.1-0.2); PAR uniformly wide, flattened, spatuliform, their base roughly even with posterior edge of ADA, without terminal spine, sigmoid, with their terminal half curving towards midline; PAR/GL = 0.8 (0.7-0.8); VS base stout, arising medially and ventrally, apex projecting dorsally, thin and curving, overall subequal in length with PAR; AP, VP, DR, transverse hinge absent.

Female

Unknown.

Other Material Examined. - None.

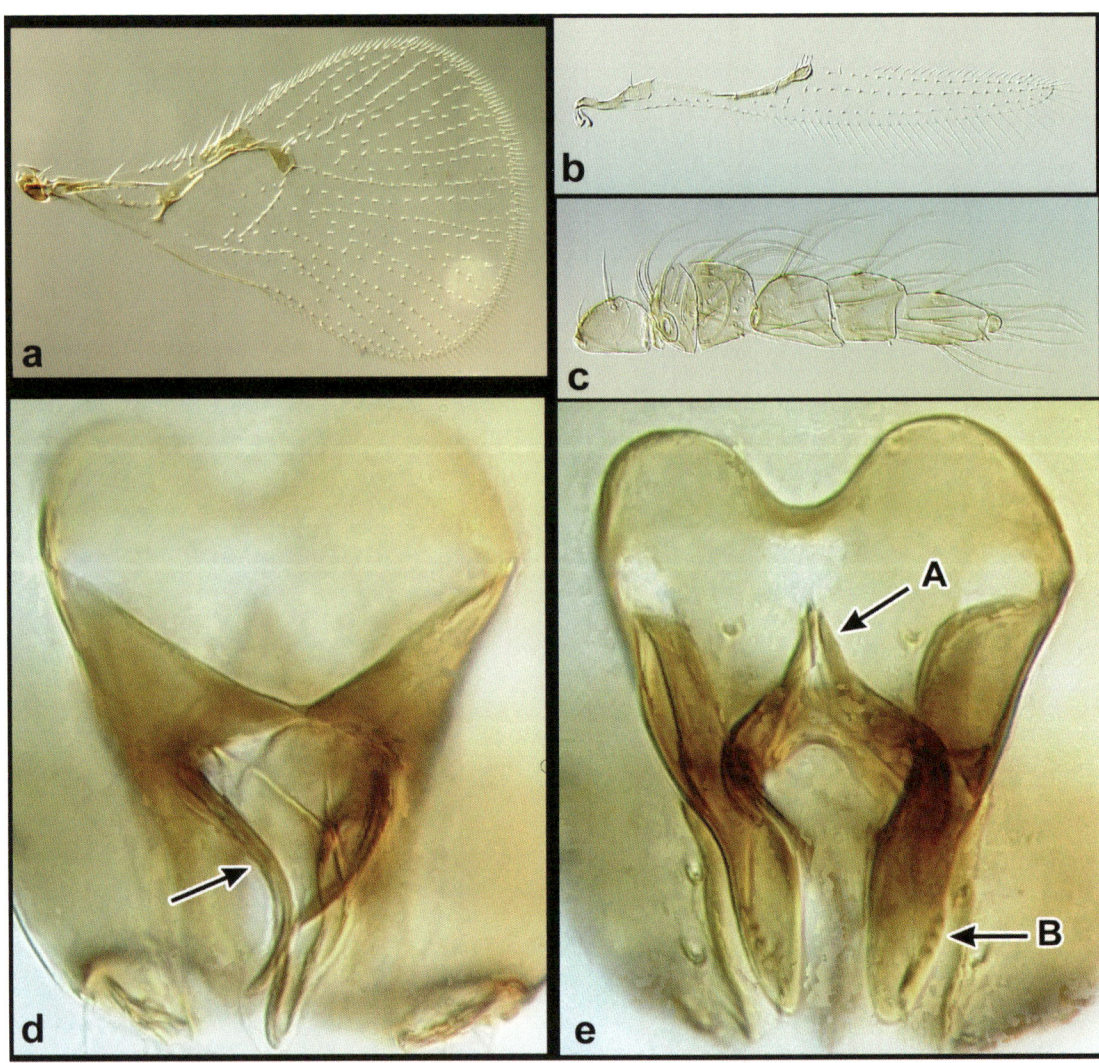

Figure 10. *Ufens acacia, ♂*. (a) forewing, dorsal; (b) hind wing, dorsal; (c) antenna, lateral; (d) genitalia, dorsal – arrow to apical portion of volsella; (e) genitalia, ventral – arrows to {A} basal portion of volsellae, {B} plate-like paramere.

Ufens acuminatus Lin, 1993
(Fig. 11)

U. acuminatus Lin, 1993: pp. 53-55.
Lin, 1994: pp. 206-207 (redescription and illustration).

Diagnosis. - Forewing densely setose with broadly diverging setal tracks r-m to M and many dispersed setae between CU1 and CU2. Hind wing width decreasing immediately apical of hamuli. Mesoscutal sculpturing longitudinally striate. Genitalia possessing parameres with terminal spine, widest near middle, the base posterior to posterior edge of anterodorsal aperture; ventral process evenly tapering; no other appendages present.

Based on wing morphology, *U. acuminatus* is likely to be confused with species such as *U. niger*, *U. principalis*, and *U. similis* which share its densely setose forewing, broadly diverging setal tracks r-m to M, and narrow hind wing. However, the genitalia of *U. acuminatus* are distinguished from these species due to their lack of volsellae and the narrow base of the ventral process. It also does not likely have the dorsal projection that these species have, though evaluation is somewhat difficult in slide-mounted specimens. Strictly in terms of genitalic similarity, this species bears a superficial resemblance to *U. lanna*. However, *U. acuminatus* has a ventral process and lacks volsellae whereas *U. lanna* has the opposite characteristics.

Types. - ◆Holotype ♀, Allotype ♂ (FACS). **CHINA**: **Fujian**: Shaxian, 20.vii.1980, N. Lin, SW; examined. Paratypes 1♂, 1♀ [not seen], same data except 20-29.vii.1980.

Distribution. - China, Thailand.

Biology. - Unknown.

Description (N=2). - BL 0.48 mm. BL/HTL = 3.6. Mesoscutal sculpturing narrowly longitudinally striate with very little interstitial sculpturing. Forewing densely setose, AA present, many dispersed setae between CU1 and CU2; FWL/HTL = 3.2 (3.1-3.2); FWL/FWW = 1.8 (1.7-1.9); FWFS/FWW = 0.09 (0.8-0.9); Max r-m to M/Min r-m to M = 5.0 (4.5-5.4); MV/PM = 1.2 (1.2-1.3); SV/MV = 0.8 (0.7-0.8); MV length/MV width = 4.0 (3.8-4.2). Hind wing width decreasing immediately apical of hamuli; HWL/HWW = 10.9; hind wing fringe long, HWFS/HWW = 0.8.

Male

Antenna: C/F = 2.1; F2/F1 = 1.1 (0.8-1.4); APB absent on funicle; 1 PLS on each of F1-C2, 2 PLS on C3; 3-4 BPS on each of F1-C1, 1 BPS on C2 and C3; 7-9 FS on F1, 9-10 FS on F2, 9-10 FS on C1, 10 FS on C2, 8-9 FS on C3, 6-7 FS on C4; US absent on each of F1-C3.

Genitalia: Capsule with basal margin somewhat transverse; GL/GW = 2.9 (2.8-3.0); GL/HTL = 0.6 (0.6-0.7); ADA/GL = 0.5 (0.4-0.5); AI absent to extremely shallow, AI/GL = 0.009 (0-0.02); PAR widest near middle, terminal spine present, their base distinctly posterior to posterior edge of ADA; PAR/GL = 0.3 (0.3-0.4); VP long and evenly tapering, base < half of capsule width, VP/GL = 0.4 (0.3-0.4); AP, VS, DR, transverse hinge absent.

Female (N=1)

Antenna: C/F = 2.1.

Ovipositor: OL/HTL = 1.2.

Other Material Examined. - **THAILAND**: Chiang Mai, 20-23.iv.1989, G. T. Baker, (1♂) (UCRC).

Comments. - Body length was indiscernible from the allotype as it was dissected. Hind wing length of the Thai specimen could not be measured as it was damaged. In the absence of rearing records, the association of the female holotype and the male allotype is questionable. Nevertheless, the non-sexually dimorphic characters are consistent between the specimens and they were collected simultaneously. Unfortunately, the holotype is a poorly cleared specimen and a determination of most antennal features was impossible.

The coordinates of the type locality according to N. Lin (pers. comm.) are 25°50'N, 117°20'E.

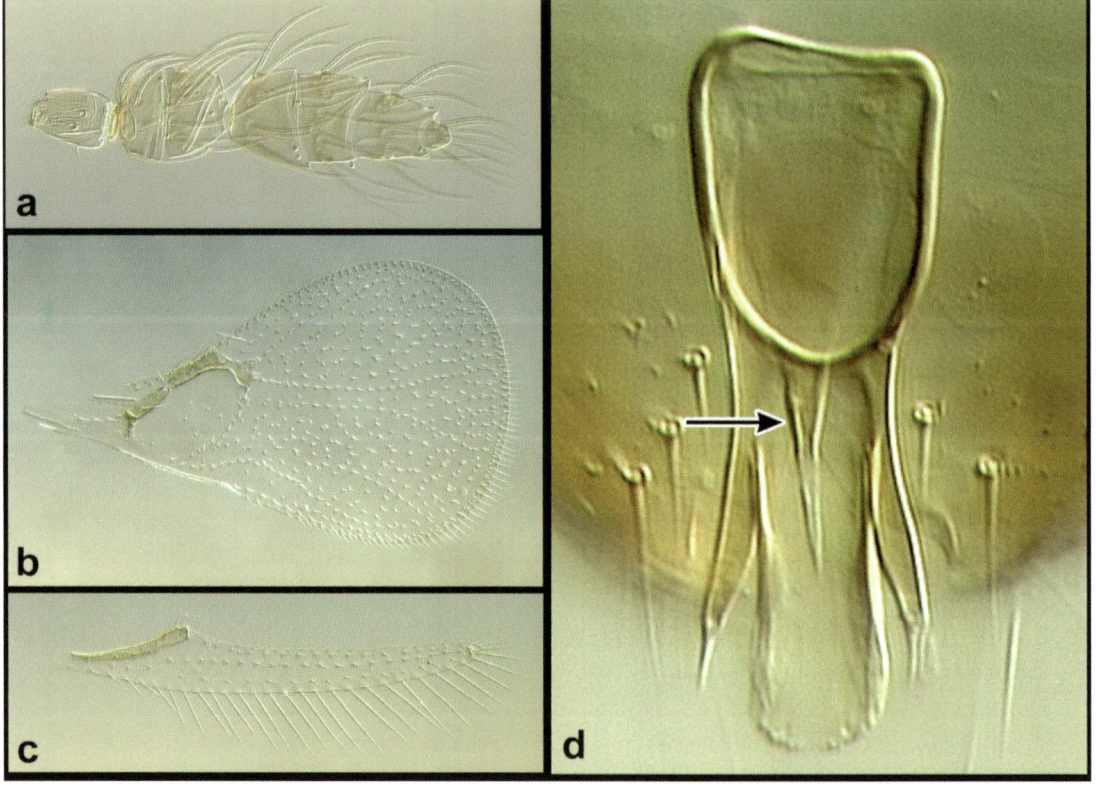

Figure 11. *Ufens acuminatus*, ♂. (a) antenna, lateral; (b) forewing, dorsal; (c) hind wing, dorsal; (d) genitalia, dorsal – arrow to ventral process.

Ufens aperserratus Owen, new species
(Fig. 12)

Diagnosis. - Forewing densely setose with narrowly diverging setal tracks r-m to M and a single setal track between CU1 and CU2. Hind wing width not decreasing immediately apical of hamuli. Mesoscutal sculpturing longitudinally striate. Genitalia with ventral process unique, its base ventral and notched, initial trajectory anterodorsal, then dramatically curving posteriorly, apical termination dorsal of rest of capsule; capsule floor fenestrate with apical longitudinal split, median sides of split serrate.

The first funicle segment, in comparison to the second, is larger in this species than in any other. However, the significance of this feature for species identification cannot be appreciated with a single specimen. The unique structures present in the capsule of this species make it highly unlikely to be confused with any other. In particular, no other species is known to have a ventral process which traverses into the interior of the capsule, terminating dorsal of the capsule itself. In addition, the longitudinal split with serrated edges (which apparently accommodates the trajectory of the ventral process) is not known to occur in any other species. The only other species known to possess appendages arising ventrally and traversing the capsule to terminate dorsally is the African *U. acacia*. However, in that case, it is the volsellae which go through the capsule; volsellae are absent in *U. aperserratus*. *U. aperserratus* also possesses an anterodorsal aperture which is distinctly longer than that of most other species.

Types. - ◆Holotype ♂ (QM). **AUSTRALIA: Northern Territory**: West of Alice Springs, Rd. to Ellery's Hole, 23°47'40"S, 133°05'20"E, 690m. el., 14.iii.2002, JMH.

Etymology. - Conjunction between the Latin apertus -open, in reference to the large ADA and serratus- saw-edged, in reference to the unique saw-like edges of the fenestrate floor of the genital capsule.

Distribution. - Australia.

Biology. - Unknown.

Description (N=1). - BL 0.71 mm. BL/HTL = 4.0. Mesoscutal sculpturing narrowly longitudinally striate without interstitial sculpturing. Forewing densely setose, AA present, single setal track between CU1 and CU2; FWL/HTL = 3.0; FWL/FWW = 1.6; FWFS/FWW = 0.05; Max r-m to M/Min r-m to M = 2.4; MV/PM = 1.0; SV/MV = 1.0; MV length/MV width = 2.3. Hind wing width not decreasing immediately apical of hamuli; Hind wing broad, HWL/HWW = 4.0; HWFS/HWW = 0.9.

Male

Antenna: C/F = 1.7; F2 short, F2/F1 = 0.6; APB absent on funicle; 6 PLS on F1, 1 PLS on each of F2-C2, 4 PLS on C3; 3 BPS on each of F1-C1, 1 BPS on C2 and C3; 1 FS on F1, 12 FS on F2, 10 FS on C1, 13 FS on C2, 9 FS on C3, 5 FS on C4; US absent on each of F1-C3.

Genitalia: Capsule broad, distinctly tapering posteriorly, anterior margin somewhat transverse; GL/GW = 2.8; GL/HTL = 1.2; ADA long, ADA/GL = 0.8; AI extremely shallow, AI/GL = 0.02; PAR with terminal spine, subequal in width along entire length, their base distinctly anterior to posterior edge of ADA; PAR/GL = 0.4; VP distinctive, base ventral and notched, trajectory initially anterodorsal, then dramatically curving posteriorly, apical termination dorsal of rest of capsule;

capsule floor fenestrate, with longitudinal split in apical 2/3, edges of split serrate; transverse hinge located near apex of capsule; AP, VS, DR absent.
<u>Female</u>
Unknown.
Other Material Examined. - None.

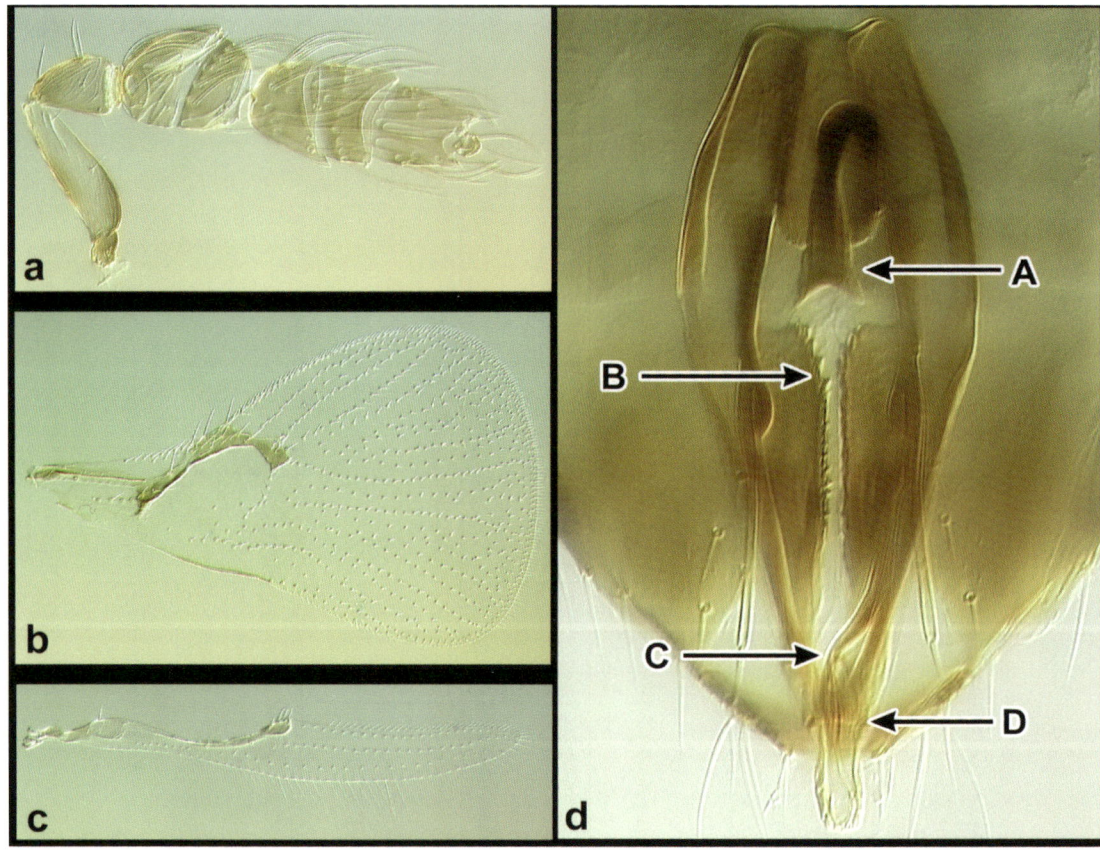

Figure 12. *Ufens aperserratus,* ♂. (a) antenna, lateral; (b) forewing, dorsal; (c) hind wing, dorsal; (d) genitalia, ventral – arrows to {A} basal portion of ventral process, {B} serrated capsule floor, {C} apical portion of ventral process, {D} transverse hinge.

***Ufens apollo* Owen, new species**
(Fig. 13)

Diagnosis. - Forewing sparsely setose, with moderately diverging setal tracks r-m to M and a single setal track between CU1 and CU2. Hind wing width decreasing immediately apical of hamuli. Mesoscutal sculpturing longitudinally striate. Genitalia with no discernible anterior invagination; parameres with terminal spine, subequal in width along entire length, base anterior to posterior edge of anterodorsal aperture; volsellae indistinguishable.

Due to their very similar genitalia, the species most likely confused with *U. apollo* are *U. niger*, *U. principalis*, *U. similis*, and *U. taniae*. *U. apollo* is the only member of this possibly allied group which has sparsely setose forewings. Also, *U. apollo* does not have genitalia with an obvious dorsal ridge, whereas *U. niger*, *U. principalis*, *U. similis* do. *U. apollo* also does not have the laterally emarginated ventral process found in *U. similis*, nor the spatulate parameres of *U. taniae*. The longitudinally striate mesoscutal sculpturing of *U. apollo* further separates it from *U. niger*, which has more distinctly cellulate sculpturing.

Types. - ◆Holotype ♂ (BMNH). **COSTA RICA: Guanacaste**: Murcielago (ACG), 75m el., 24.i-4.ii.1996, J. Ugalde, MT.

Etymology. - Named for Apollo, the ancient Greek god of music, healing, light, and truth.

Distribution. - Costa Rica, México.

Biology. - Unknown.

Description (N=3). - BL 0.6 (0.4-0.7) mm. BL/HTL = 3.4 (3.2-3.6). Mesoscutal sculpturing longitudinally striate to nearly longitudinally cellulate with interstitial sculpturing primarily transverse. Forewing sparsely setose, AA present, single setal track between CU1 and CU2; FWL/HTL = 3.1 (3.0-3.3); FWL/FWW = 1.6; FWFS/FWW = 0.08 (0.07-0.08); Max r-m to M/Min r-m to M = 3.3 (2.8-4.2); MV/PM = 1.1 (1.0-1.2); SV/MV = 0.9 (0.8-1.0); MV length/MV width = 3.3 (2.8-3.7). Hind wing width decreasing immediately apical of hamuli; HWL/HWW = 8.5 (8.1-8.9); HWFS/HWW = 1.0 (0.9-1.0).

Male

Antenna: Club comparatively long, C/F = 2.8 (2.6-3.1); F2/F1 = 1.4 (1.2-1.6); APB absent on funicle; 1 PLS on each of F1-C2, 2 PLS on C3; 4-5 BPS on each of F1-C1, 1 BPS on C2 and C3; 7-10 FS on F1, 8-10 FS on F2, 8-9 FS on C1, 8-10 FS on C2, 7-8 FS on C3, 6 FS on C4; US absent on each of F1-C3.

Genitalia: Capsule gradually tapering from base; GL/GW = 2.7 (2.6-2.8); GL/HTL = 1.0 (0.9-1.0); ADA/GL = 0.6; PAR with terminal spine, subequal in width along entire length, their base distinctly anterior to posterior edge of ADA; PAR/GL = 0.3; VP width at base > half of capsule width, evenly tapering, with indistinct ring at base, VP/GL = 0.4 (0.3-0.4); dorsal projection present; VS indistinguishable; AP, AI, DR, transverse hinge absent.

Female (N=1)

Antenna: C/F = 2.6; F2/F1 = 1.2; 1 APB on F1 and F2, 1 APB on C3; 1 PLS on F1, 2 PLS on each of F2-C2, 4 PLS on C3; 5-6 BPS on each of F1-C1, 1 BPS on C2 and C3; 0 FS on F1 and F2, 6 FS on C1, 9 FS on C2, 3 FS on C3; 1 UPP on C3; 10-12 US on each of F1-C1, 0 US on C2, 4 US on C3.

Ovipositor: OL/HTL = 0.9.

Other Material Examined. - **COSTA RICA: San José**: Cuidad Colón, 800m. el., iii.iv.1990, L. Fournier and P. Hanson (1♂) (USNM). **MÉXICO: Yucatán**: Chichén Itzá, 27.vii.1984, G. Gordh, SW (1♂, 1♀) (UCRC).

Comments. - Based upon comparisons with species possessing similar genitalia such as *U. similis*, it seems likely that *U. apollo* does have volsellae and a dorsal projection. However, they cannot be clearly distinguished in any of the specimens available. The area of genitalia posterior of the anterodorsal aperture (and likely including the dorsal projection) is dramatically bent in all slide mounts except for the holotype.

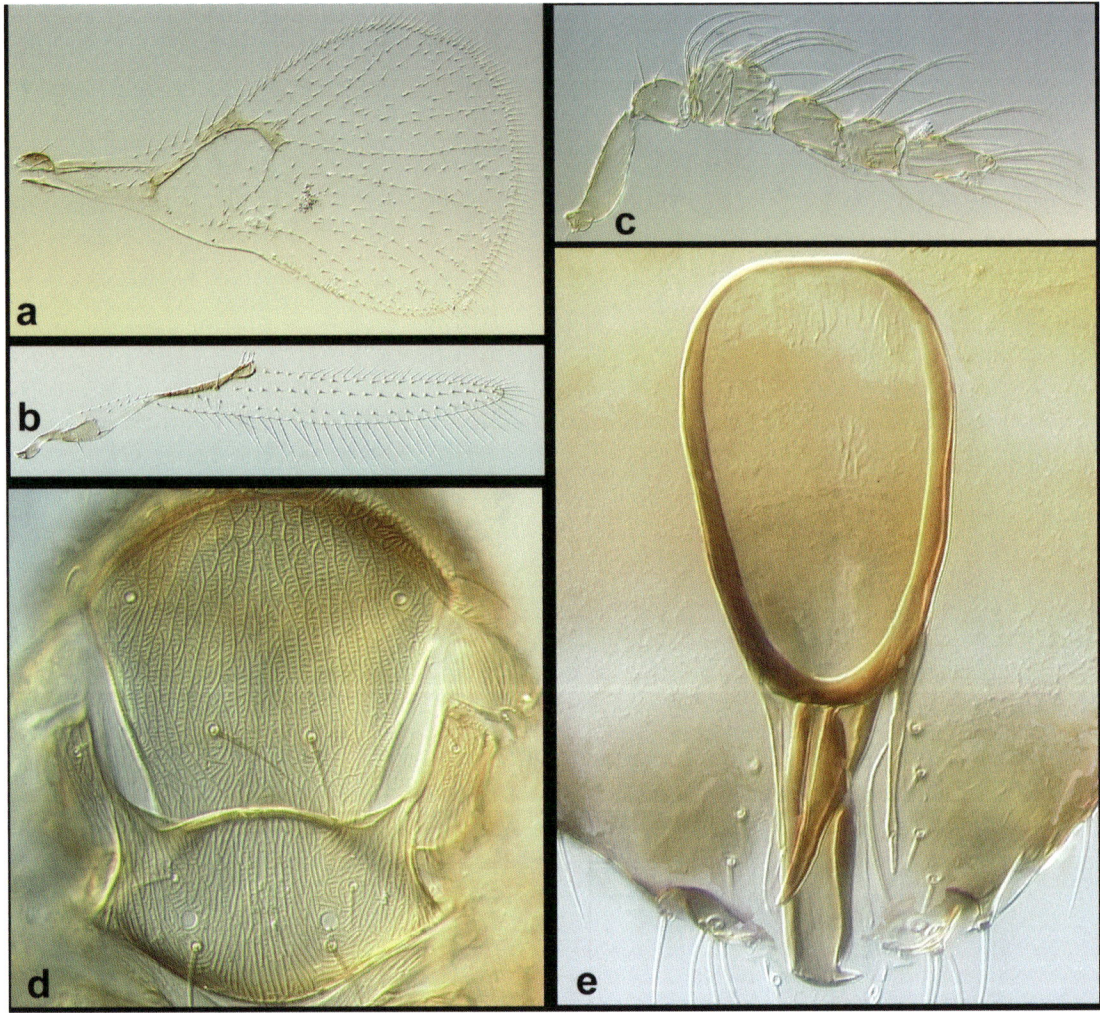

Figure 13. *Ufens apollo,* ♂. (a) forewing, dorsal; (b) hind wing, dorsal; (c) antenna, lateral; (d) mesosoma, dorsal; (e) genitalia, dorsal.

Ufens austini **Owen, new species**
(Fig. 14)

Diagnosis. - Forewing densely setose, with narrowly diverging setal tracks r-m to M and a single setal track between CU1 and CU2. Hind wing width not decreasing immediately apical of hamuli. Mesoscutal sculpturing longitudinally cellulate. Genitalia with anterodorsal aperture abruptly constricted at ca. half its length; parameres with terminal spine, subequal in width along entire length, base anterior to posterior edge of anterodorsal aperture; dorsal ridge present; a distinctly bulbous area occurring immediately posterior of transverse hinge.

U. austini may most likely be confused with *U. australensis*, or less likely with *U. nazgul*. It differs from *U. australensis* by the parameres being narrower at their base, presence of a dorsal ridge, and a distinct bulbous region posterior to the transverse hinge. *U. austini* can be differentiated from *U. nazgul* by its more compact club segments, more densely setose forewing disk, lack of alar acanthae, and long, thin parameres with a terminal spine. In *U. nazgul* the parameres are without a terminal spine, widest near middle, and dorsoventrally flattened.

Types. - ◆Holotype ♂ (QM). **AUSTRALIA: Southern Australia**: Adelaide Hills, Cox Scrub Conservation Park, 35°20'01"S, 138°44'56"E, 27.xii.2003 - 17.i.2004, A. D. Austin, MT. Paratypes 2♂, 1♀, same data (UCRC, except 1♂ paratype ANIC).

Etymology. - Named for A. D. Austin, collector of most of the known specimens of this species.

Distribution. - Australia.

Biology. - Unknown.

Description (N=4). - BL 0.7 (0.6-0.8) mm. BL/HTL = 3.8 (3.6-4.1). Mesoscutal sculpturing longitudinally striate to nearly longitudinally cellulate with interstitial sculpturing lightly rugulose. Forewing densely setose; AA present; single setal track between CU1 and CU2; FWL/HTL = 3.1 (2.8-3.6); FWL/FWW = 1.6; FWFS/FWW = 0.03; Max r-m to M/Min r-m to M = 2.2 (1.3-2.9); MV/PM = 1.0 (0.8-1.0); SV/MV = 1.0; MV length/MV width = 2.1 (2.0-2.1). Hind wing width not decreasing immediately apical of hamuli; HWL/HWW = 7.7 (7.3-8.1); HWFS/HWW = 0.8 (0.7-0.8).

Male

Antenna: Club segments compact; C/F = 2.0 (1.8-2.3); F2/F1 = 1.0 (0.8-1.2); APB absent on funicle; 1 PLS on each of F1-C2, 2 PLS on C3; 3-5 BPS on each of F1-C1, 1 BPS on C2 and C3; 8-10 FS on F1, 10-14 FS on F2, 11-12 FS on C1, 13-14 FS on C2, 10-11 FS on C3, 6-8 FS on C4; US absent on each of F1-C3.

Genitalia: Capsule gradually tapering, heavily sclerotized ventral portion of capsule laterally extending from base of paramere to posterior of transverse hinge (i. e. entire length of parameres); GL/GW = 3.1 (2.8-3.4); GL/HTL = 1.2 (1.2-1.4); ADA abruptly constricted in posterior half, ADA/GL = 0.5; AI shallow, AI/GL = 0.4 (0.3-0.7); PAR with terminal spine, subequal in width along entire length, their base distinctly anterior to posterior edge of ADA; PAR/GL = 0.4; VP base likely hollow and < half of capsule width, somewhat wider at base then evenly tapering, VP/GL = 0.5; capsule with a transverse hinge located at ca. posterior third, followed by a distinctly bulbous area, then long, parallel-sided to apex; DR present; AP, VS absent.

Female (N=1)

Antenna: C/F = 2.0; F2/F1 = 1.1; 1 APB on F1 and F2, 0 APB on C3; 2 PLS on F1, 3 PLS on F2, 1 PLS on C1 and C2, 4 PLS on C3; 3-4 BPS on each of F1-C1, 1 BPS on C2 and C3; 0 FS on F1 and F2, 6 FS on C1, 11 FS on C2, 6 FS on C3; 1 UPP on C3; 3-5 US on each of F1-C1, 0 US on C2 and C3.

Ovipositor: OL/HTL = 1.7.

Other Material Examined. - **AUSTRALIA: Southern Australia**: Brachina Gorge, 31°20'S, 138°34'E, xi.1987, I. Naumann and J. Cardale, YPT (1♂) (ANIC).

Comments. - The only other species known in which the female antenna lacks unsocketed setae on C2 and C3 is *U. pallidus*. It is questionable if the lack of unsocketed setae is significant. Several specimens of *U. pallidus* were examined, but only one *U. austini* was studied. Unsocketed setae can sometimes be difficult to discern in slide-mounted specimens, so their absence on C2 and C3 in the latter species may be an artifact. Other traits do not suggest a relationship between these species.

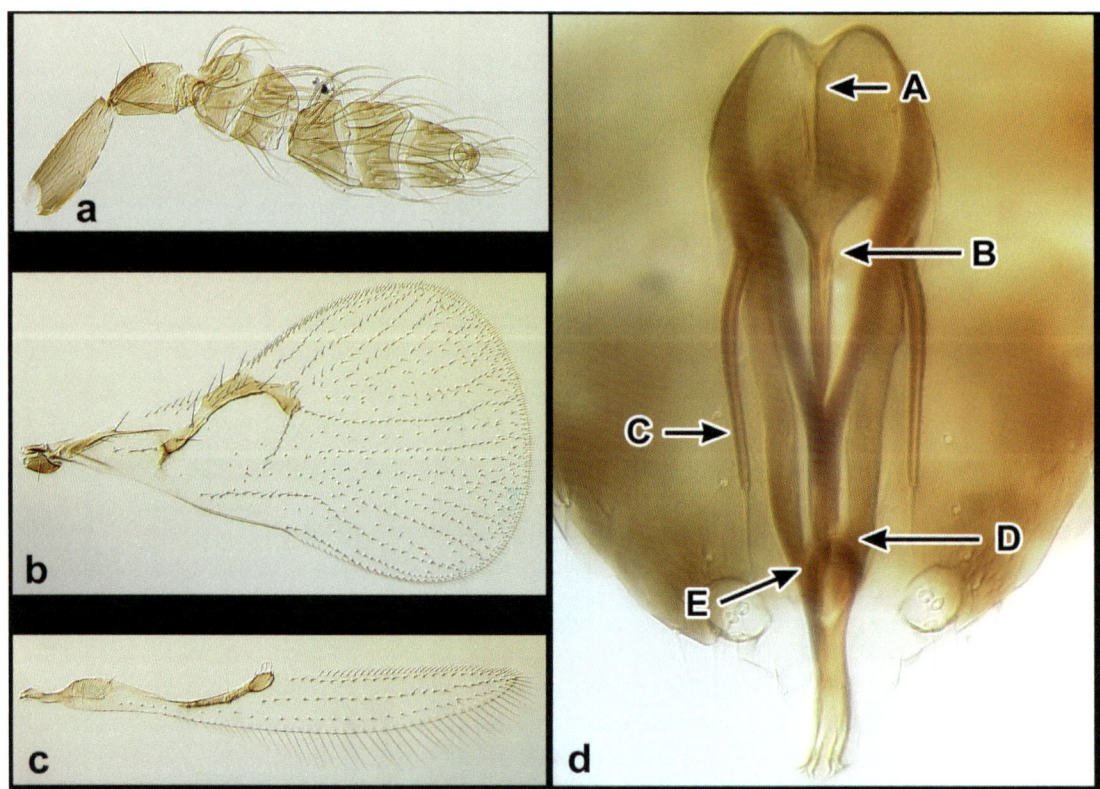

Figure 14. *Ufens austini,* ♂. (a) antenna, lateral; (b) forewing, dorsal; (c) hind wing, dorsal; (d) genitalia, dorsal – arrows to {A} dorsal ridge, {B} basal portion of ventral process, {C} parameres, {D} transverse hinge, {E} bulbous area.

Ufens australensis Owen, new species
(Fig. 15)

Diagnosis. - Forewing densely setose, with narrowly diverging setal tracks r-m to M and a single setal track between CU1 and CU2. Hind wing width not decreasing immediately apical of hamuli. Mesoscutal sculpturing longitudinally striate. Genitalia with apodemes absent; parameres with terminal spine, widest at base, base anterior to posterior edge of anterodorsal aperture; transverse hinge in posterior quarter of capsule; without bulbous area posterior of transverse hinge.

 U. australensis is most easily confused with *U. austini*, but can be separated by the somewhat greater basal width of the parameres, as well as lack of a distinct constriction in the anterodorsal aperture, and absence of both a dorsal ridge and a bulbous region posterior to the transverse hinge.

Types. - ◆Holotype ♂ (ANIC). **AUSTRALIA: Southern Australia**: Brookfield Conservation Park, 34°24'S, 139°30'E, 19.ii.1992, J. C. Cardale, on / near *Melaleuca* flowers. Paratype 2♂; 1♂ same data (ANIC), 1♂ same data except 34°21'S, 139°29'E, 18.ii.1992 (UCRC).

Etymology. - Named for the country of origin.

Distribution. - Australia.

Biology. - Unknown.

Description. - BL 0.7 (0.7-0.8) mm. BL/HTL = 4.1 (3.5-5.2). Mesoscutal sculpturing narrowly longitudinally striate with interstitial sculpturing lightly rugulose. Forewing sparsely setose; AA present; single setal track between CU1 and CU2; FWL/HTL = 3.3 (3.1-3.9); FWL/FWW = 1.5; FWFS/FWW = 0.04 (0.03-0.04); Max r-m to M/Min r-m to M = 2.2 (1.4-2.7); MV/PM = 1.0 (0.9-1.1); SV/MV = 0.9 (0.8-1.1); MV length/MV width = 2.0 (1.8-2.2). Hind wing width not decreasing immediately apical of hamuli; HWL/HWW = 6.9 (6.4-7.4); HWFS/HWW = 0.8 (0.7-0.9).

Male

Antenna: Club segments compact; C/F = 2.3 (1.8-3.0); F2/F1 = 1.2 (0.9-1.7); APB absent on funicle; 1 PLS on each of F1-C2, 2 PLS on C3; 2-4 BPS on each of F1-C1, 1 BPS on C2, 0-1 BPS on C3; 9-11 FS on F1, 12-15 FS on F2, 10-13 FS on C1, 12-14 FS on C2, 9-11 FS on C3, 5-8 FS on C4; US absent on each of F1-C3.

Genitalia: Capsule gradually tapering; anterolateral margin gradually curving; GL/GW = 3.2 (3.1-3.4); GL/HTL = 1.2 (1.0-1.4); ADA abruptly constricted at posterior half, ADA/GL = 0.6 (0.5-0.7); AI shallow, AI/GL = 0.3 (0.1-0.4); PAR with terminal spine, tapering and wider at base, their base distinctly anterior to posterior edge of ADA; PAR/GL = 0.5 (0.5); VP elongate, base likely hollow, < half of capsule width, somewhat wider at base then evenly tapering, VP/GL = 0.6 (0.6-0.7); transverse hinge located ca. posterior 1/4 of capsule, not followed by distinctly bulbous area; AP, DR, VS absent.

Female

Unknown.

Other Material Examined. - **AUSTRALIA: Southern Australia**: Brachina Creek, 31°20'S, 138°33'E, 9.xi.1987, I. Naumann and J. Cardale, ex. ethanol (1♂); Dingly

Dell Camp, Oraparinna Crk., 31°21'S, 138°42'E, 7.xi.1987, I. Nauman and J. Cardale, ex. ethanol (1♂); Brookfield Conservation Park, 34°19'S, 139°30'E, 2.xii.1991 - 2.i.1992, J. Stelman and S. Williams, MT, mallee with *Triodia* (1♂). **Queensland**: 13 km SE Weipa, 12°40'S, 143°00'E, 16.i - 16.ii.1994, P. Zborowski and D. Khalu (1♂). **Western Australia**: Nambung NP, Hangover Bay, 30°35.54'S, 115°06.44'E, 13-14.xi.2002, Hawks, JG, JBM, and AKO, YPT (1♂).

Comments. - Molecular data for *U. australensis* were presented in Owen et al. (2007) as *Ufens* sp. 11, and can be found under Genbank accession numbers AY623540 (28S-D2+D3) and AY940381 (18S).

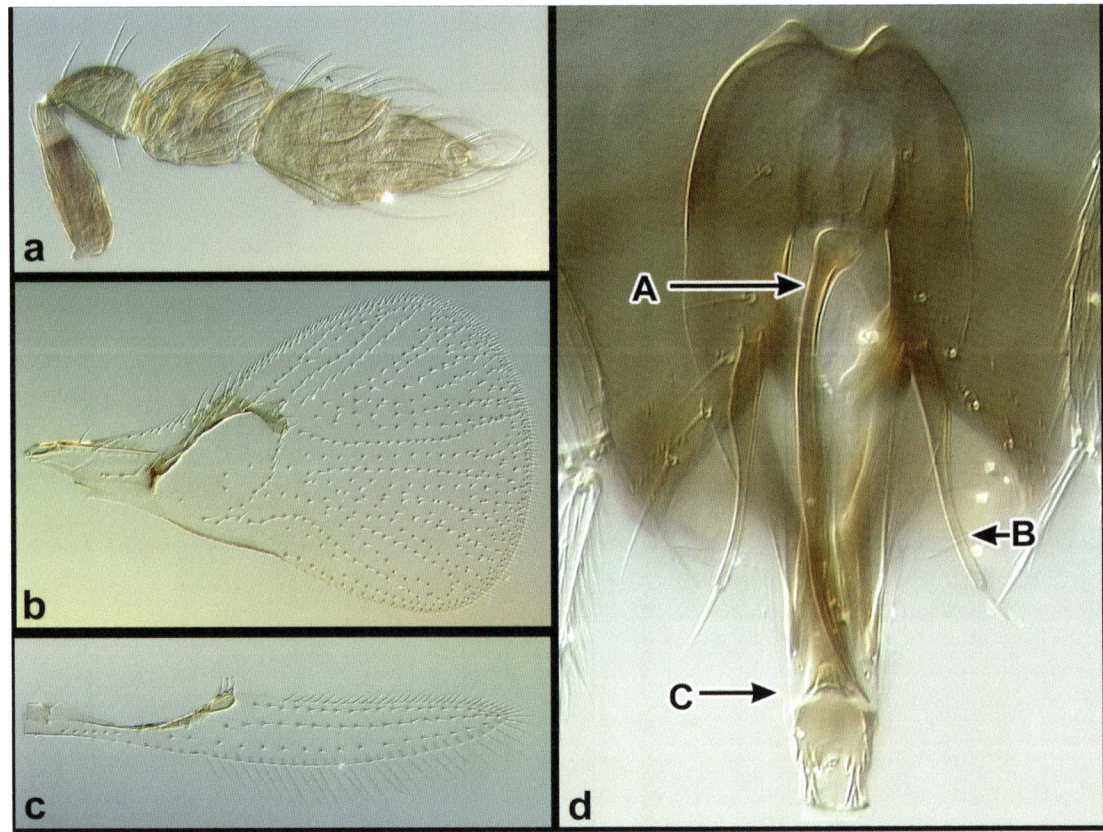

Figure 15. *Ufens australensis,* ♂. (a) antenna, lateral; (b) forewing, dorsal; (c) hind wing, dorsal; (d) genitalia, dorsal – arrows to {A} basal portion of ventral process, {B} paramere, {C} transverse hinge.

Ufens bestiolis Owen, new species
(Fig. 16)

Diagnosis. - Forewing sparsely setose, with narrowly diverging setal tracks r-m to M and a single setal track between CU1 and CU2. Hind wing width not decreasing immediately apical of hamuli. Mesoscutal sculpturing longitudinally striate. Genitalia with anterodorsal aperture 'V' shaped; parameres with terminal spine, subequal in width along entire length, base even with posterior edge of anterodorsal aperture; volsellae with minute sclerotized hook at end; dorsal ridge present.

    U. bestiolis is most likely to be confused with *U. parvimalis*, but it can be distinguished by its single placoid sensillum on the first funicle segment, more longitudinally striate mesoscutal sculpturing, and genitalia with dorsal ridge present and a shorter anterodorsal aperture.

Types. - ◆Holotype ♂ (QM). **AUSTRALIA: Queensland**: 9 km E of Blackbutt, Blackbutt Crk., 22.ix.1995, JDP, SW. Paratype ♂, same data (UCRC).

Etymology. - Bestiola –ae, Latin for a small animal.

Distribution. - Australia.

Biology. - Unknown.

Description. - BL 0.6 (0.6-0.7) mm. BL/HTL = 3.7 (3.4-3.9). Mesoscutal sculpturing longitudinally striate with interstitial sculpturing lightly rugulose. Forewing sparsely setose; AA present; single setal track between CU1 and CU2; FWL/HTL = 3.0 (2.9-3.2); FWL/FWW = 1.6 (1.6-1.7); FWFS/FWW = 0.06 (0.05-0.08); Max r-m to M/Min r-m to M = 1.8 (1.6-2.1); MV/PM = 1.1 (1.0-1.2); SV/MV = 0.9 (0.8-1.0); MV length/MV width = 2.5 (1.5-3.0). Hind wing width not decreasing immediately apical of hamuli; HWL/HWW = 8.0 (7.6-8.4); HWFS/HWW = 0.9 (0.9-1.0).

Male

Antenna: Club segments compact; C/F = 2.0 (1.8-2.1); F2/F1 = 1.2 (0.9-1.4); APB absent on funicle; 1 PLS on each of F1-C2, 2 PLS on C3; 1-5 BPS on each of F1-C1, 1 BPS on C2 and C3; 6-10 FS on F1, 8-14 FS on F2, 8-13 FS on C1, 8-14 FS on C2, 7-10 FS on C3, 5-6 FS on C4; US absent on each of F1-C3.

Genitalia: Capsule somewhat quadrate; GL/GW = 2.9 (2.7-3.1); GL/HTL = 0.8 (0.7-1.0); posterior margin of ADA 'V' shaped, ADA/GL = 0.5; AI shallow, AI/GL = 0.07 (0.05-0.09); PAR with terminal spine, subequal in width along entire length, their base even with posterior edge of ADA; PAR/GL = 0.4; VP base < half of capsule width, VP/GL = 0.4 (0.3-0.4); transverse hinge located ca. posterior third of capsule; VS apparently with minute sclerotized hook at end, VS/GL = 0.3; DR present; AP absent.

Female

Unknown.

Other Material Examined. - **AUSTRALIA: Australian Capital Territory**: Canberra, Jerrabomberra Wetlands NR, 35°18.8'S, 149°09.7'E, 27.xi.2002, JDP, SW (1♂). **Northern Territory**: Standley Chasm, W of Alice Springs, 23°43'32"S, 133°26'10"E, 720m. el., 12.iii.2002, JMH, SW *Eucalyptus* (1♂); Ormiston Gorge, W of Alice Springs, 23°39'11"S, 132°43'37"E, 13.iii.2002, JMH, SW dry savanna

(1♂). **Queensland**: Auburn River NP, Auburn Falls area, 151°04'E, 25°44'S, 23.ix.1995, JDP, SW 1° *Callistemon* (4♂); Biggenden, 32 km SE, Munna Crk., 24.ix.1995, JDP, SW (4♂); 62 km E Mt. Isa to Cloncurry, 322 m el., 20°47'18"S, 139°04'51"E, 7.iii.2002, JMH, SW savanna scrub (1♂). **Western Australia**: Erskine Conservation Park, 32°34'S, 115°41'E, 27.xii.1999, C. J. Burwell, SW heath (1♂); Mount Jetty Crk. floodplain, Munbinea Rd., 30°32.53'S, 115°13.53'E, 12-13.xi.2002, JG, D. Hawks, JBM, and AKO, YPT (1♂).

Comments. - Molecular data for *U. bestiolis* were presented in Owen et al. (2007) as *Ufens* sp. 5, and can be found under Genbank accession numbers AY623531 (28S-D2+D3) and AY940373 (18S).

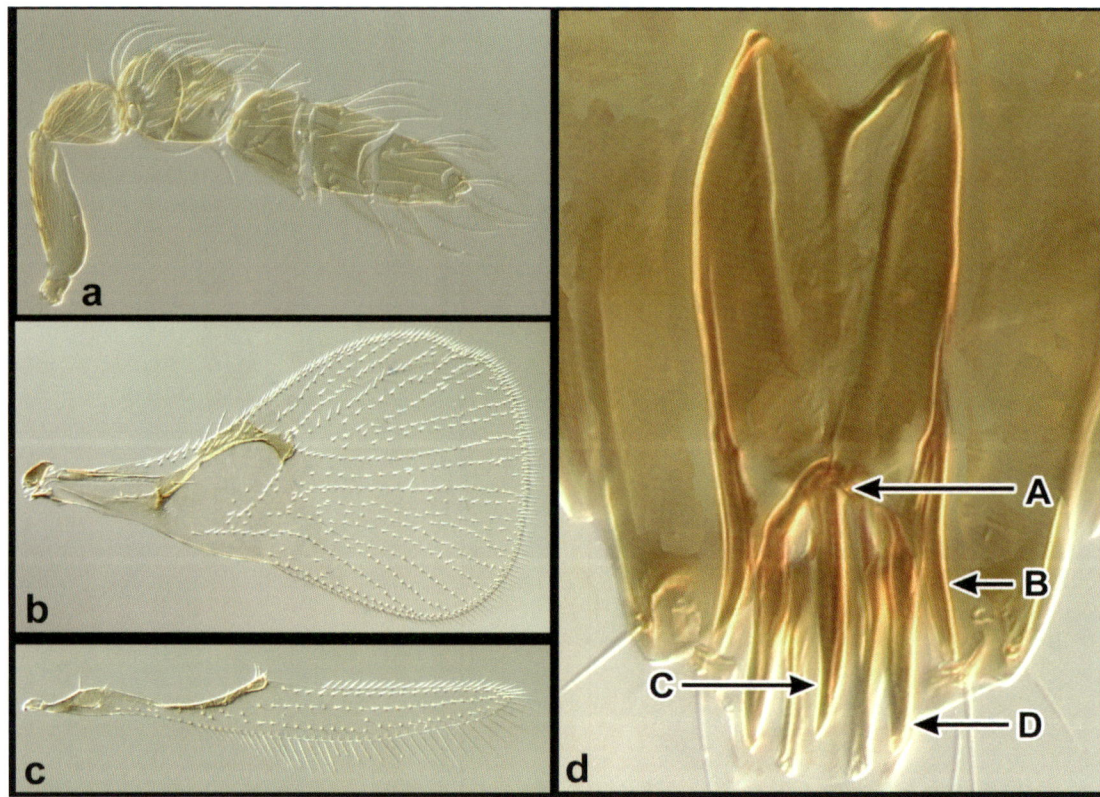

Figure 16. *Ufens bestiolis,* ♂. (a) antenna, lateral; (b) forewing, dorsal; (c) hind wing, dorsal; (d) genitalia, dorsal – arrows to {A} transverse hinge, {B} paramere, {C} ventral process, {D} volsella.

Ufens cardalia Owen, new species
(Fig. 17)

Diagnosis. - Antenna with abundant placoid sensilla on the second funicular and second club segments. Forewing sparsely setose; a single setal track between CU1 and CU2. Hind wing width not decreasing immediately apical of hamuli. Mesoscutal sculpturing longitudinally striate to narrowly cellulate. Genitalia with a narrow capsule; apodemes present; parameres with terminal spine, subequal in width along entire length, their base even with posterior edge of anterodorsal aperture; volsellae absent; dorsal ridge absent.

U. cardalia is most likely to be confused with *U. kurrajong* and *U. pintoi*, both of which also have aedeagal apodemes and somewhat similar capsule shapes. However, *U. cardalia* can be distinguished by a narrower capsule and shorter parameres than either of those species. It is further differentiated from *U. kurrajong* by the presence of a distinct anterior invagination in the genital capsule, and from *U. pintoi* by the lack of volsellae and an anterodorsal aperture which is not abruptly constricted posteriorly.

Types. - ◆Holotype ♂ (ANIC). **AUSTRALIA: Northern Territory**: 30 km W of Alice Springs, 23°24'S, 133°50'E, 8.v.1978, J. C. Cardale, ex. ethanol.

Etymology. - A derivative of Cardale, collector of the only known specimen.

Distribution. - Australia.

Biology. - Unknown.

Description (N=1). - BL 0.9 mm. BL/HTL = 4.0. Mesoscutal sculpturing longitudinally striate to nearly longitudinally cellulate with interstitial sculpturing transverse to rugulose. Forewing sparsely setose; AA absent; single setal track between CU1 and CU2; FWL/HTL = 3.0; FWL/FWW = 1.4; FWFS/FWW = 0.03; MV/PM = 0.9; SV/MV = 1.0; MV length/MV width = 2.2. Hind wing width does not decrease immediately apical of hamuli; HWL/HWW = 7.2; HWFS/HWW = 0.6.

Male

Antenna: C/F = 2.0; F2 distinctly longer than F1, F2/F1 = 1.8; APB absent on funicle; 1 PLS on F1, 5 PLS on F2, 1 PLS on C1, 4 PLS on C2, 2 PLS on C3; 1-2 BPS on each of F1-C1, 1 BPS on C2 and C3; 10 FS on F1, 12 FS on F2, 13 FS on C1, 13 FS on C2, 11 FS on C3, 7 FS on C4; US absent on each of F1-C3.

Genitalia: Capsule narrow; GL/GW = 4.3; GL/HTL = 1.0; ADA/GL = 0.5; AI narrow but distinct, AI/GL = 0.12; PAR with terminal spine, subequal in width along entire length, their base even with posterior edge of ADA, PAR short, PAR/GL = 0.2; VP width at base > half of capsule width, wider at base then evenly tapering, VP/GL = 0.3; transverse hinge located at ca. posterior half of capsule, parallel sided posterior of transverse hinge; AP present; DR, VS absent.

Female

Unknown.

Other Material Examined. - None.

Comments. - The separated forewing of the holotype is damaged, with the apex torn off, thereby prohibiting the calculation of Max r-m to M/Min r-m to M. Setal patterns of both fore and hind wings are difficult to discern as they are overly cleared and nearly transparent. The second funicle segment is considerably longer than the first in the holotype. However, with only a single specimen, intraspecific variation cannot be appreciated.

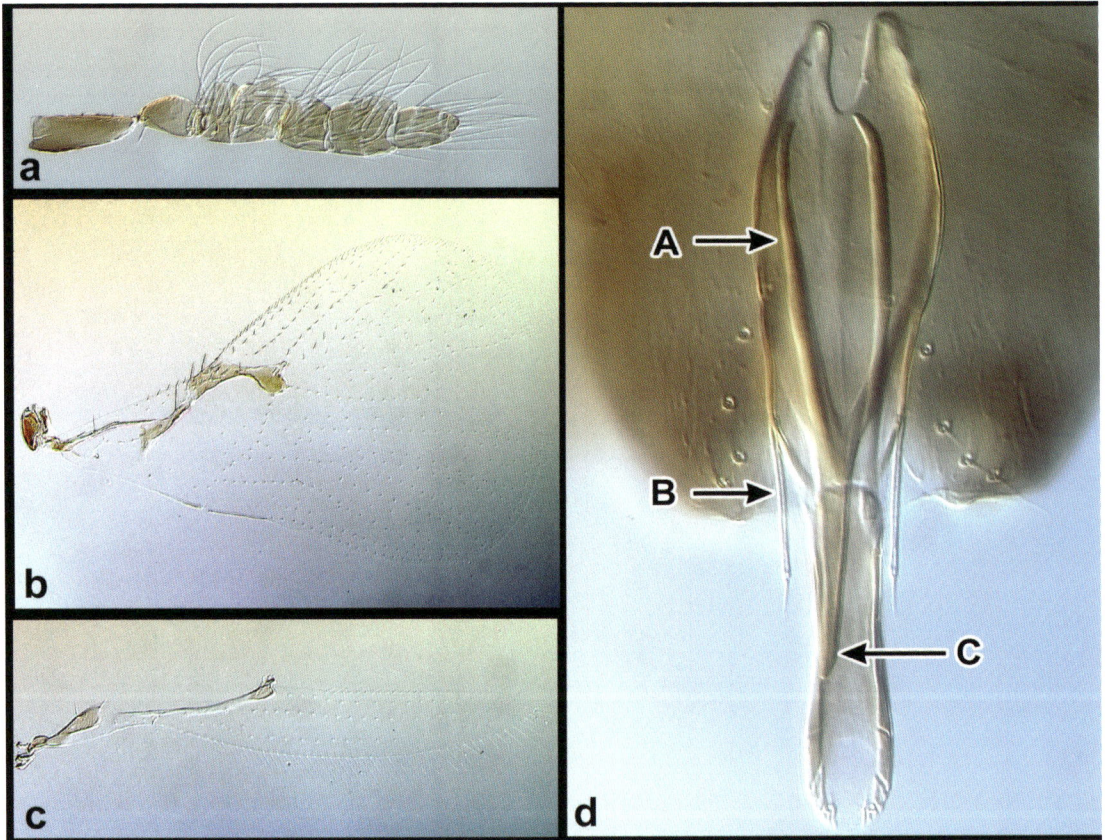

Figure 17. *Ufens cardalia,* ♂. (a) antenna, lateral; (b) forewing, dorsal; (c) hind wing, dorsal; (d) genitalia, dorsal – arrows to {A} apodeme, {B} paramere, {C} apex of ventral process.

Ufens ceratus **Owen, 2005**
(Fig. 18)
Ufens sp. Triapitsyn et al., 2002: pp. 40-41; Triapitsyn, 2003: pp. 253-254.
Ufens ceratus Owen, in Al-Wahaibi et al., 2005: pp. 276-280.

Diagnosis. - Body generally with marked sexual dimorphism in color; females dark brown, males light yellow. Head of males with a set of forward-projecting stout setae on the lateral margins of the clypeus and two adjacent pairs on the genae. Forewing sparsely setose, with narrowly diverging setal tracks r-m to M and a single setal track between CU1 and CU2. Hind wing width not decreasing immediately apical of hamuli. Mesoscutal sculpturing longitudinally striate. Genitalia with anterior invagination pronounced; parameres with terminal spine, subequal in width along entire length, their base posterior posterior edge of anterodorsal aperture; volsellae filiform; dorsal ridge absent.

   This is one of the most distinctive *Ufens* species, as no other is known with males possessing stout setae on the gena. Additionally, no other is known with similar levels of color dimorphism generally present. However, although not reported in Al-Wahaibi et al. (2005), some males of this species are known to be as darkly pigmented as females. The genitalia are also distinguishing, with only the Australian *U. invaginatus* possessing a similarly pronounced anterior invagination. *U. ceratus* is separated, however, by its parameres being subequal along their length, and lack of a ventral process. The females of this species have among the greatest numbers of placoid sensilla on their second funicle segment. Other species whose females have many placoid sensilla on F2 include *U. debachi* and *U. pallidus*. However, no other features suggest a relationship to these species.

Types. - ◆Holotype ♂, Allotype ♀ (USNM). **UNITED STATES: California**: <u>San Bernardino County</u>, Crafton Hills, 13.iii.2003, L. Higgins, ex. glassy-winged sharpshooter eggs on citrus leaves. Paratypes 8♂, 5♀, same data; 2♂, 2♀ card mounted (1♂, 1♀ BMNH, remainder UCRC).

Distribution. México, United States.

Biology. *U. ceratus* has been reared from *Oncometopia clarior* (Walker) (Hemiptera: Cicadellidae) ['?' on original label], from *H. coagulata* on citrus (cf. Triapitsyn et al. 2002, 2003; Al-Wahaibi et al. 2005), *Simmondsia chinensis* (Link) Schneid. (jojoba), and *Cercis* sp. (redbud) (Fabaceae), and from *H. liturata* on *S. chinensis* and *Cassia* sp. (Fabaceae). It is also know from undetermined leafhopper hosts on *Vitus* sp. (grape) (Vitaceae) and *Ulmus* sp. (elm) (Ulmaceae), and from undetermined hosts on *Hyptis* sp. (Lamiaceae) and *S. chinensis*.

Description. - Color generally sexually dimorphic (see above). BL 0.7 (0.5-0.9) mm. BL/HTL = 3.3 (2.8-3.7). Mesoscutal sculpturing longitudinally striate with interstitial sculpturing transverse. Forewing sparsely setose; AA absent; single setal track between CU1 and CU2; FWL/HTL = 2.7 (2.2-3.2); FWL/FWW = 1.6 (1.5-1.6); FWFS/FWW = 0.10 (0.08-0.10); Max r-m to M/Min r-m to M = 1.7 (1.1-2.1); MV/PM = 1.1 (0.9-1.2); SV/MV = 1.0 (0.9-1.2); MV length/MV width = 2.6 (2.0-3.2). Hind wing width not decreasing immediately apical of hamuli; HWL/HWW = 7.6 (6.6-8.6); HWFS/HWW = 0.9 (0.8-1.0).

<u>Male</u>
Color generally primarily yellow with midlobe of mesoscutum brown medially; scutellum brown; metanotum brown laterally; gena generally yellowish, but sometimes darker than the rest of head. Head with 6 forward-projecting stout setae: 1 pair on clypeus near lateral margins; 2 pairs adjacent to clypeus on genae (not equally stout in all specimens but always distinct from surrounding setae).
Antenna: Club segments somewhat loosely joined; C/F = 2.2 (2.0-2.4); F2/F1 = 1.2 (1.0-1.6); APB absent on funicle; 1 PLS on each of F1-C2, 2 PLS on C3; 2-5 BPS on each of F1-C1, 1 BPS on C2 and C3; 6-11 FS on F1, 9-16 FS on F2, 8-12 FS on C1, 8-14 FS on C2, 7-12 FS on C3, 5-8 FS on C4; US absent on each of F1-C3.
Genitalia: GL/GW = 2.0 (1.7-2.4); GL/HTL = 0.7 (0.6-0.8); ADA/GL = 0.5 (0.5); AI extremely pronounced, AI/GL = 0.3 (0.2-0.3); PAR with terminal spine, subequal width along entire length, their base posterior to posterior edge of ADA; PAR/GL = 0.4 (0.3-0.4); VS filiform, arising medially, sometimes separated for ca. half of their length before becoming appressed apically, VS/GL = 0.5 (0.5-0.6); dorsal projection present; transverse hinge somewhat rounded and immediately posterior of ADA; VP, DR, AP absent.

<u>Female</u>
Color typically dark brown with dorsal portion of head brownish-orange; genae dark brown and distinctly darker than the rest of head; antennae tan; femora yellow or banded dark brown and yellow; tibiae and tarsi yellow; midlobe of mesoscutum light brown medially; scutellum light brown; metanotum yellow medially.
Antenna: C/F = 1.7 (1.6-1.8); F2/F1 = 1.8 (1.2-2.9); 1 APB on F1 and F2, 1 APB on C3; 1 PLS on F1, 5-10 PLS on F2, 1-2 PLS on C1 and C2, 4 PLS on C3; 2-6 BPS on each of F1-C1, 1 BPS on C2 and C3; 0 FS on F1 and F2, 9-10 FS on C1, 11-13 FS on C2, 2-3 FS on C3; 1 UPP on C3; 6-11 US on F1 and F2, 3-7 US on C1, 0 US on C2, 0-2 US on C3.
Ovipositor: OL/HTL = 1.0 (0.9-1.2).

Other Material Examined. - **UNITED STATES**: **Arizona**: <u>*Pinal County*</u>: 7 mi. W Superior: 2,350 ft., 26.iv.1980, S. Manweiler, ex. eggs inserted in leaf of *S. chinensis*, Jojoba project 291042 (3♂, 4♀); 2,350 ft., 9.v.1980, S. Manweiler, on *S. chinensis*, Jojoba project 290366 (1♀); 2,500 ft., 25.v.1980, em. 10.vi.1980 from *S. chinensis*, Univ Calif Insect Survey specimen #290386 (1♂); 2,500 ft., 25.v.1980, em. 10.vi.1980 from *S. chinensis*, University of California Insect Survey specimen #290386 (1♂); 2,500 ft., 21.vi.1980, S. A. Manweiler, beaten from *S. chinensis*, Jojoba project 291158 (1♀); 2,350 ft., 4.x.1980, S. Manweiler, on *S. chinensis*, em. i.1981, Jojoba project 291328 (1♂, 1♀). **California**: <u>*Imperial County*</u>: Salton City, ex. smoketree sharpshooter egg mass on *Cassia* seed pod, via R. S. Mendés (1♂, 4♀; 3♀ card mounted). <u>*Riverside County*</u>: Coachella, "Old Shop", 12.vii.1989, D. Gonzalez, ex. grape leaves with leafhopper eggs (1♂); Deep Canyon, unknown host on *Hyptis*, 23.iii.1963 (1♂, 6♀); Indio, 30.x.1986, D. Goodward coll., ex. leafhopper eggs on Siberian elm leaf (2♂, 4♀); Palm Desert, 21.viii/17.ix.1986, D. Goodward, ex leafhopper eggs on elm (4♂, 6♀; 2♂, 2♀ card mounted); Riverside, UCR Agricultural Experiment Station, Field 7E, 3.vii.2000, A. K. Al-Wahaibi, ex. *Homalodisca* sp. on *S. chinensis* (2♂, 2♀); 5.6 mi. S Sage on R3, Sec. 32 T.75, R.IE.

site 2, 116°54'W, 33°31'N, 24-29.ii.1980, Jojoba project #303378, 303388 (2♂, 1♀); 5.6 mi. S Sage on R3, 29.ii1980, Sec. 32 T.75, R.IE. site 2, 116°54'W, 33°31'N, em. 4.iv.1980, S. Frommer, ex. eggs on *S. chinensis*, Jojoba project #302019 (1♀). *San Diego County*: San Diego, 26.vii.1972, Powers, ex. *Homalodisca lacerta* (1♂, 5♀) [1 slide]. **Florida**: *Jefferson County*: Monticello, ARC, ex. *H. coagulata* eggs on redbud, vi-viii.1979, J. C. Ball (1♂, 11♀). **Texas**: *Hidalgo County*: Bentsen Rio Grande State Park, 19.vi.1986, JBW, SW, River Hiking Trail (1♂); *Presidio County*: Big Bend Ranch SNA, Yedra Canyon, 20.vi.1991, JBW, 91/029 (1♂); Big Bend Ranch SNA, Agua Adentro, 18/23.vi.1991, R. Wharton (1♂). **MÉXICO**: **Baja California Sur**: La Paz, 10 km W, 28.x.1983, JDP, SW (1♂, 1♀). **Nuevo Leon**: Allende, 6 km S, roadside of Hwy. 85, Sanatorio Naturista de Canoas, ex. *Oncometopia clarior* (Walker) ['?' on original label] eggs on orange, 10.iv.2000, L. Bezark and S. Triapitsyn, SandR # 00-07-01 (1♂, 1♀). **Tamaulipas**: Llera de Canales, 8.iii.2000, ex. sharpshooter egg mass on orange leaf in private garden, S. Triapitsyn (3♂, 3♀).

Comments. – Molecular data for *U. ceratus*, as presented in Owen et al. (2007), can be found under Genbank accession numbers AY623533 (28S-D2+D3) and AY940375 (18S).

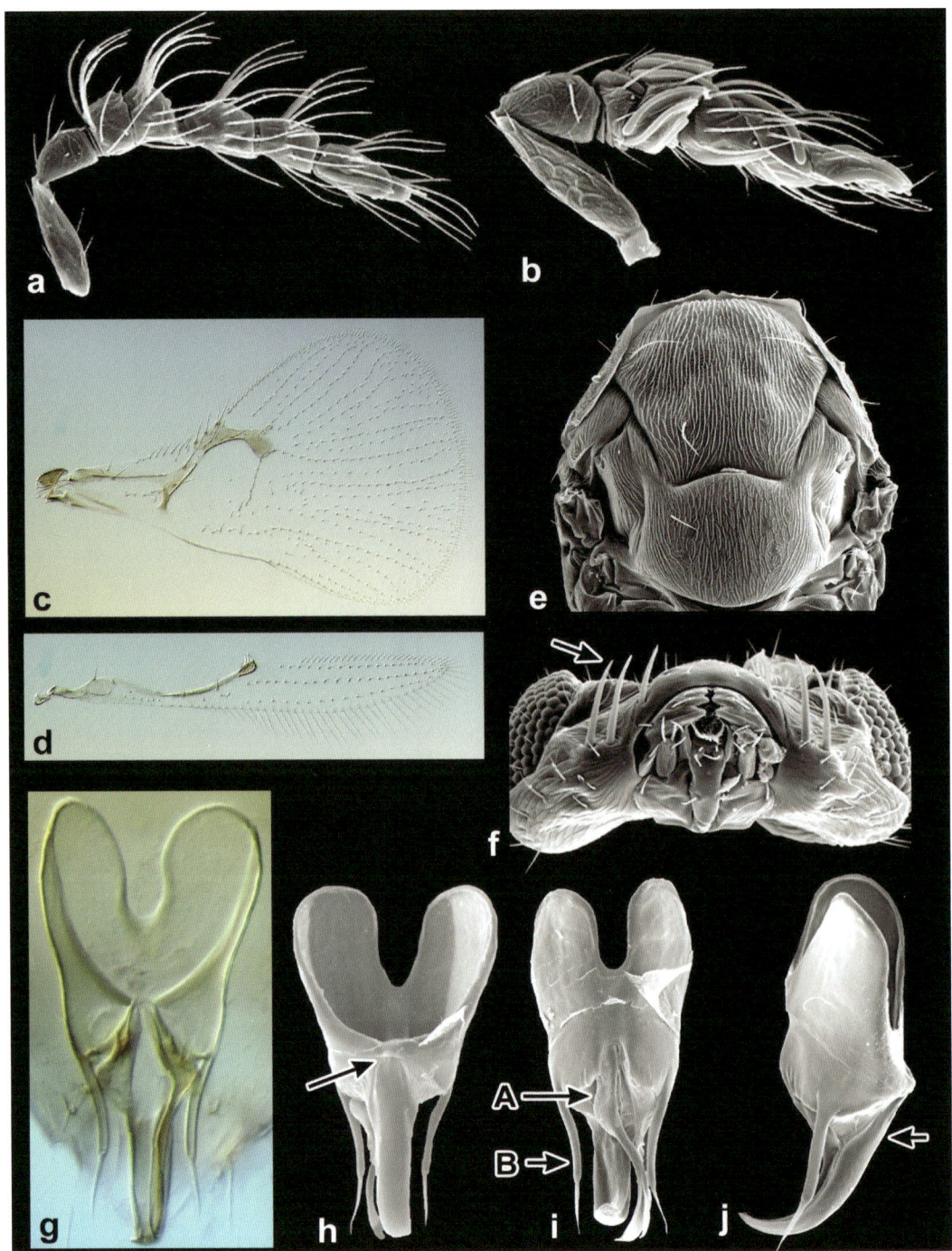

Figure 18. *Ufens ceratus*. (a) ♂ antenna, medial; (b) ♀ antenna, medial; (c) forewing, dorsal; (d) hind wing, dorsal; (e) mesosoma, dorsal; (f) head, ventral – arrow to stout setae; (g) ♂ genitalia, dorsal; (h) ♂ genitalia, dorsal – arrow to transverse hinge; (i) ♂ genitalia, ventral – arrows to {A} base of volsella, {B} paramere; (j) ♂ genitalia, lateral – arrow to dorsal projection.

Ufens cupuliformis Lin, 1993
(Fig. 19)

U. cupuliformis Lin, 1993: pp. 55-56.
Lin, 1994: pp. 210-211 (redescription and illustration).

Diagnosis. - Forewing sparsely setose with narrowly diverging setal tracks r-m to M and a single setal track between CU1 and CU2. Hind wing width not decreasing immediately apical of hamuli. Mesoscutal sculpturing longitudinally striate. Genitalia with anterior margin of capsule broadly convex; parameres laterally emarginate near tip and apically diverging from midline, lacking a terminal spine, their base even with posterior edge of anterodorsal aperture; volsellae absent.

This species is most likely to be confused with other species that also have parameres without a terminal spine and which apically diverge from the midline, such as *U. khamai* and *U. mezentius*. *U. cupuliformis* can be separated from both of those species, however, by its parameres that are laterally emarginated and relatively short, as well as by its broadly convex anterior margin.

Types. - Holotype ♀, Allotype ♂ (FACS). **CHINA: Fujian**: Songxi, 25°52'N, 117°19'E, 23.viii.1987, Z. Wu, SW. Paratype 1 ♀, 1 ♂, as above except: ♀ Yongchun, 25°32'N, 118°10'E, 8.ix.1987, X. Chen, SW; ♂, Xianyou, 25°31'N, 118°11'E, 29.ix.1987, Z. Wu, SW. (FACS)

Distribution. - China.

Biology. - Unknown.

Description (N=1). - BL 0.6 mm. BL/HTL = 3.6. Mesoscutal sculpturing very narrowly longitudinally striate without interstitial sculpturing. Forewing sparsely setose; AA absent; single setal track between CU1 and CU2; FWL/HTL = 3.0; FWL/FWW = 1.6; FWFS/FWW = 0.07; Max r-m to M/Min r-m to M = 1.7; MV/PM = 1.1; SV/MV = 0.9; MV length/MV width = 3.0. Hind wing width does not decrease immediately apical of hamuli; Hind wing broad, HWL/HWW = 4.4; HWFS/HWW = 0.9.

Male

Antenna: Club segments somewhat loosely joined; club comparatively long, C/F = 3.2; F2/F1 = 1.3; APB absent on funicle; 1 PLS on each of F1-C2, 2 PLS on C3; 4 BPS on each of F1-C1, 1 BPS on C2 and C3; 7 FS on F1, 8 FS on F2, 7 FS on C1, 8 FS on C2, 10 FS on C3, 5 FS on C4; US absent on each of F1-C3.

Genitalia: Capsule broad anteriorly, anterior margin broadly convex; GL/GW = 1.5; GL/HTL = 0.9; ADA/GL = 0.3; PAR widest at base, base approximately even with posterior edge of ADA, without terminal spine, laterally emarginate near tip, and apically diverging from midline; PAR/GL = 0.4; transverse hinge, AI, VP, AP, VS, DR absent.

Female (N=1)

Antenna: C/F = 2.2; F2/F1 = 1.0; 1 APB on F1 and F2, 0 APB on C3; 1-2 PLS on each of F1-C2, 4 PLS on C3; 3 BPS on each of F1-C1, 1 BPS on C2 and C3; 0 FS on F1 and F2, 6-7 FS on C1 and C2, 3 FS on C3; 1 UPP on C3; 4-6 US on each of F1-C1, 0 US on C2, 3 US on C3.

Ovipositor: OL/HTL = 1.3.

Material Examined. - **CHINA: Fujian**: Shaxian, 10.vii.1981, N. Lin (1♂); Yongchun, 8.ix.1985, N. Lin (1♀).

Comments. - The slides of the specimens examined were unclear, so all measurements may be prone to more error than typical of other species. The anterodorsal aperture of the male genitalia was estimated in spite of the rip in the capsule. The specimens examined were identified by Lin as *U. cupuliformis*, and they clearly conform to the original description of this species (Lin 1993)

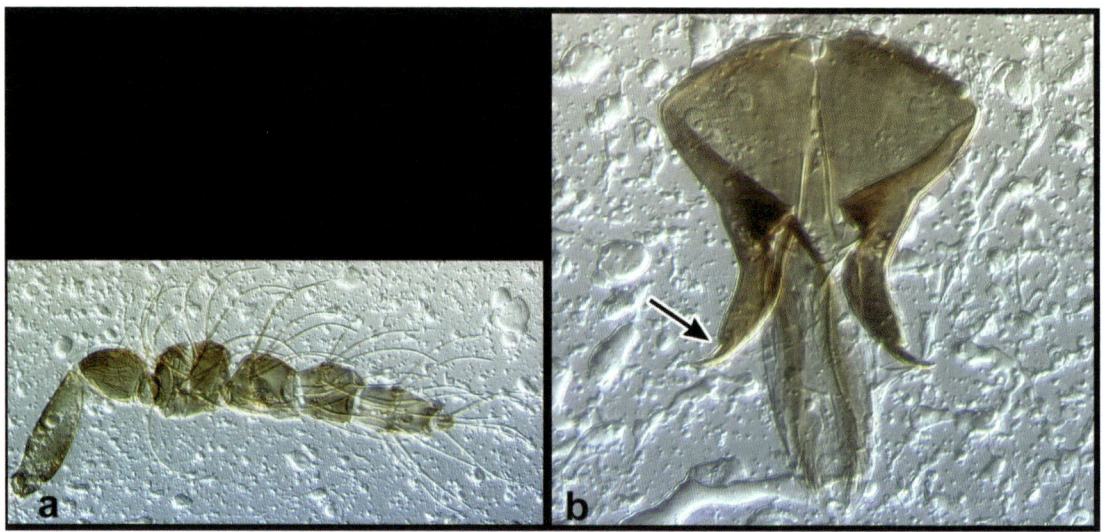

Figure 19. *Ufens cupuliformis,* ♂. (a) antenna, lateral; (b) genitalia, dorsal – arrow to lateral emargination of parameres.

Ufens debachi Owen, new species
(Fig. 20)

Diagnosis. - Forewing densely setose with narrowly diverging setal tracks r-m to M and a single setal track between CU1 and CU2. Hind wing width increasing apical of hamuli. Mesoscutal sculpturing longitudinally striate. Genitalia possessing stout parameres without terminal spine but with heavily sclerotized tips, their base even with posterior edge of anterodorsal aperture; volsellae filiform.

This species is most likely to be confused with the Palearctic *U. forcipis*, as both possess long, straight parameres without a terminal spine. However, *U. debachi* can be differentiated by the presence of volsellae, the gradually tapering genital capsule, and the much longer anterodorsal aperture.

Types. - ♦Holotype ♂, Allotype ♀ (USNM). **MÉXICO: Baja California Sur**: Santiago (Las Barracas), 1-11.v.1989, P. Debach, YPT. Paratype ♂, same data (UCRC).

Etymology. - Named for Paul Debach, the collector of most of the known material of this species.

Distribution. - México, United States.

Biology. - Unknown.

Description. - BL 0.7 (0.7-0.8) mm. BL/HTL = 3.4 (3.2-3.7). Mesoscutal sculpturing longitudinally striate with interstitial sculpturing transverse. Forewing fairly densely setose; AA absent; single setal track between CU1 and CU2; FWL/HTL = 2.7 (2.6-2.8); FWL/FWW = 1.4 (1.4-1.5); FWFS/FWW = 0.04; Max r-m to M/Min r-m to M = 1.6 (1.5-1.9); MV/PM = 0.8 (0.6-0.9); SV/MV = 1.5 (1.2-1.9); MV length/MV width = 1.8 (1.6-2.0). Hind wing width increasing beyond hamuli; HWL/HWW = 6.9 (6.5-7.2); HWFS/HWW = 0.6 (0.5-0.7).

Male

Antenna: Club segments somewhat loosely joined; C/F = 2.2 (2.1-2.3); F2/F1 = 1.0 (0.8-1.5); APB absent on funicle; 1 PLS on each of F1-C2, 2 PLS on C3; 6-8 BPS on each of F1-C1, 1 BPS on C2, 1-2 BPS on C3; 10-12 FS on F1, 13-17 FS on F2, 13-16 FS on C1, 13-17 FS on C2, 11-12 FS on C3, 6-9 FS on C4; US absent on each of F1-C3.

Genitalia: GL/GW = 3.1 (3.0-3.3); GL/HTL = 1.0; ADA/GL = 0.6; AI slight, AI/GL = 0.3 (0.3-0.4); PAR without terminal spine, somewhat wider at base and gradually tapering, tips apparently heavily sclerotized, their base even with posterior edge of ADA; PAR/GL = 0.3 (0.2-0.4); VS filiform, subequal in length to PAR; transverse hinge immediately posterior of termination of ADA; DR, AP, VP absent.

Female (N=3)

Antenna: C/F = 2.0 (1.8-2.4); F2/F1 = 0.9 (0.8-1.0); 1 APB on F1 and F2, 0 APB on C3; 3-4 PLS on F1, 7-10 PLS on F2, 4-5 PLS on C1, 3-4 PLS on C2, 4-5 PLS on C3; 5-8 BPS on each of F1-C1, 1 BPS on C2 and C3; 0 FS on F1 and F2, 10-11 FS on C1, 13-17 FS on C2, 2-3 FS on C3; 1 UPP on C3; 6-9 US on F1, 4-5 US on F2, 3 US on C1, 0 US on C2, 2-4 US on C3.

Ovipositor: OL/HTL = 1.1 (1.1-1.2).

Other Material Examined. - **UNITED STATES**: **Arizona**: <u>Cochise Co.</u>: Dragoon Mtns., Stronghold, 12-16.viii.1970, R. J. Shaw, UV trap (1♂); <u>Pima Co.</u>: Brawley Wash, 2500' el., 3.viii.1982, G. Gibson (1♂, 1♀). **MÉXICO**: **Baja California Sur**: Santiago, 30km E (Las Barracas), iv-vi.1984-1989, P. Debach, YPT (7♂, 8♀).

Comments. - The genitalia in the SEM mount (Fig. 20 f, g) is extremely bent at the transverse hinge, lending credence to the hypothesis that it does indeed function as a hinge. The darkened tips of the parameres evident in all slide-mounted material are not obvious in SEM micrographs. It is assumed that these tips are more heavily sclerotized than the rest of the capsule, as they have been seen protruding in card-mounted specimens and do indeed appear dark.

The only *Ufens* species currently known from Baja California are *U. ceratus*, *U. debachi*, and *U. simplipenis*, though other Nearctic species would likely be found with an increased sampling effort. Nevertheless, of the species known, only *U. debachi* has been collected in any numbers from this region. This bias is almost certainly due to the concentrated collecting effort in Las Barracas by Paul Debach. The only other species collected there, *U. simplipenis*, has been confirmed from only a single specimen. In contrast, although Arizona and the southwestern United States in general have been comparatively well collected, *U. debachi* is represented there by only two individuals. This suggests that *U. debachi* is a rare component of the *Ufens* fauna of the southwestern United States, but is the dominant *Ufens* in at least some areas of Baja California Sur. Further collecting efforts are needed to more fully evaluate this hypothesis.

The antenna of females of this species are notable for having a greater than normal numbers of placoid sensilla on the first funicle segment, and flagelliform seta on the first and second club segments.

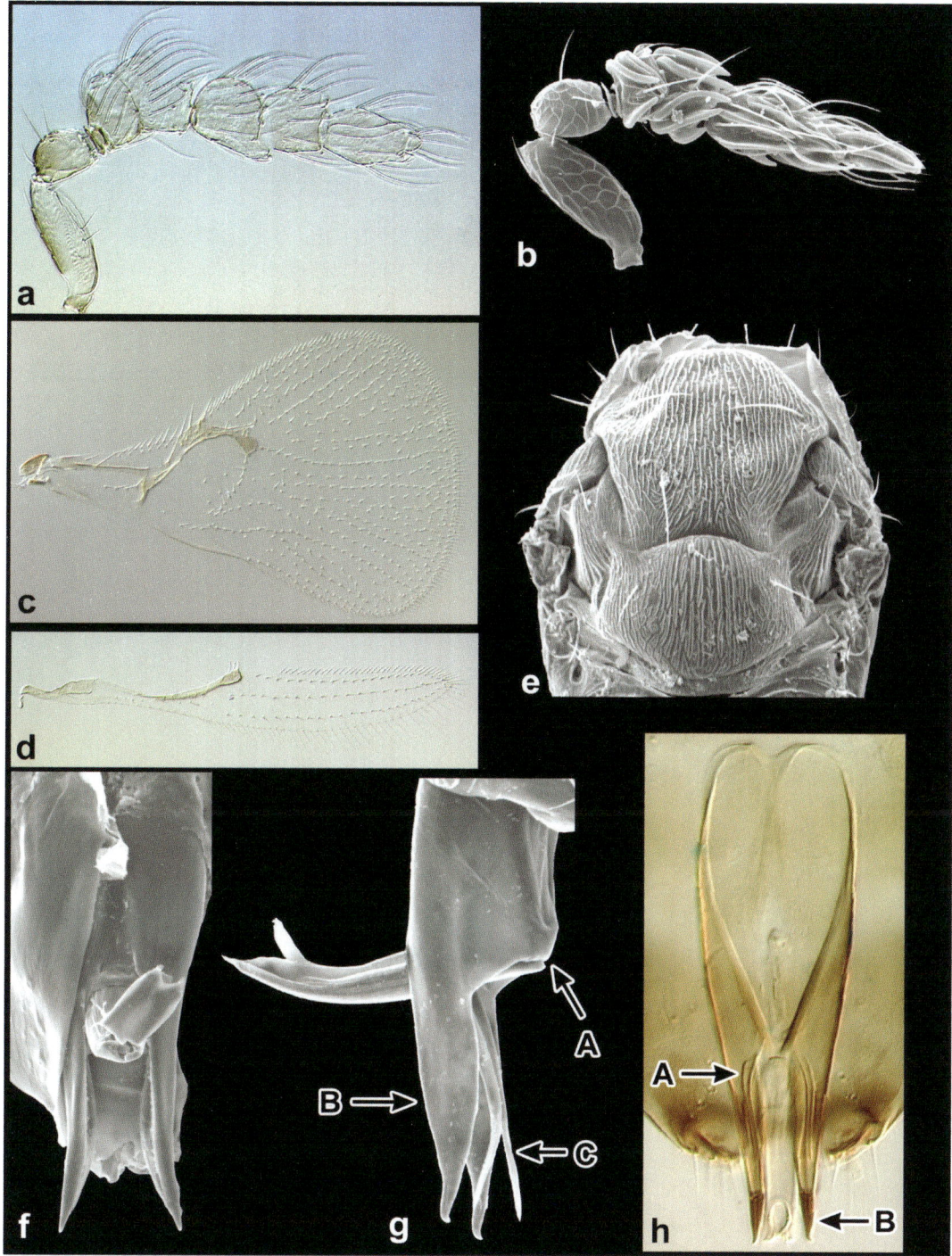

Figure 20. *Ufens debachi*. (a) ♂ antenna, lateral; (b) ♀ antenna, lateral; (c) forewing, dorsal; (d) hind wing, dorsal; (e) mesosoma, dorsal; (f) ♂ genitalia, ventral; (g) ♂ genitalia, lateral – arrows to {A} transverse hinge, {B} paramere, {C} volsella; (h) ♂ genitalia, dorsal – arrows to {A} base of volsella, {B} darkened apex of paramere.

Ufens decipiens Owen, new species
(Fig. 21)

Diagnosis. - Forewing sparsely setose with narrowly diverging setal tracks r-m to M and a single setal track between CU1 and CU2. Hind wing width not decreasing immediately distal to hamuli. Mesoscutal sculpturing longitudinally cellulate (reticulate?). Genitalia with parameres short and with a terminal spine, their base anterior to posterior edge of anterodorsal aperture; volsellae highly modified, sickle-shaped, strongly curved toward midline.

This species has distinctive genitalia, which separate it from all other species except *U. gloriosus. U. decipiens* can be differentiated, however, by its lack of alar acanthae, presence of unsocketed setae on the funicle, presence of flagelliform setae on F1, the single placoid sensillum on the C1 (vs. 4-6 in *U. gloriosus*), and the two placoid sensilla on C3 (vs. 4-7 in *U. gloriosus*).

Types. - ♦Holotype ♂ (ANIC). **AUSTRALIA**: **South Australia**: Brachina Creek, 31°20'S, 138°33'E, 9.xi.1987, I. Naumann and J. Cardale, ex. ethanol.

Etymology. - Latin for deceiving, as in a species closely resembling another; in reference to the close resemblance of the genitalia of this species and those of *U. gloriosus.*

Distribution. - Australia.

Biology. - Unknown.

Description (N=2). - BL 0.9 mm. BL/HTL = 4.7 (4.7). Mesoscutal sculpturing longitudinally cellulate with interstitial sculpturing primarily transverse. Forewing sparsely setose; AA present; single setal track between CU1 and CU2; FWL/HTL = 3.0 (2.9-3.0); FWL/FWW = 1.7 (1.6-1.7); FWFS/FWW = 0.04; Max r-m to M/Min r-m to M = 1.4 (1.4-1.5); MV/PM = 1.0 (1.0-1.1); SV/MV = 0.8; MV length/MV width = 2.1 (2.0-2.2). Hind wing width not decreasing immediately apical of hamuli; HWL/HWW = 7.7 (7.4-8.0); HWFS/HWW = 0.8.

Male

Antenna: Club compact; C/F = 1.8 (1.7-1.9); F2/F1 = 1.1 (1.0-1.2); 1-2 APB on F1, 0 APB on F2; 3-4 PLS on F1, 2 PLS on F2, 1 PLS on C1, 2 PLS on C2 and C3; 3-6 BPS on each of F1-C1, 1 BPS on C2 and C3; 3 FS on F1, 13 FS on F2, 14 FS on C1, 12-13 FS on C2, 10-11 FS on C3, 7 FS on C4; 0-2 US on F1, 1-3 US on F2. 0-1 US on C1, US absent on C2 and C3.

Genitalia: Capsule nearly parallel sided in anterior half, anterior margin nearly straight; GL/GW = 3.7; GL/HTL = 1.3 (1.2-1.3); ADA/GL = 0.6; AI extremely shallow, AI/GL = 0.03; PAR with terminal spine, subequal in width along entire length, their base distinctly anterior to posterior edge of ADA; PAR/GL = 0.2; VS unusual, somewhat sickle-shaped with apex curved in towards midline; AP length/GL = 0.4; transverse hinge in apical quarter of genitalia; DR present, extending beyond quarter of GL; VP base < half maximum width of genital capsule, VP/GL = 0.3.

Female

Unknown.

Other Material Examined. - **AUSTRALIA**: **Australian Capital Territory**:
Canberra, Black Mtn., 8-13.iii.1999, G. Gibson, YPT (1♂) (UCRC).

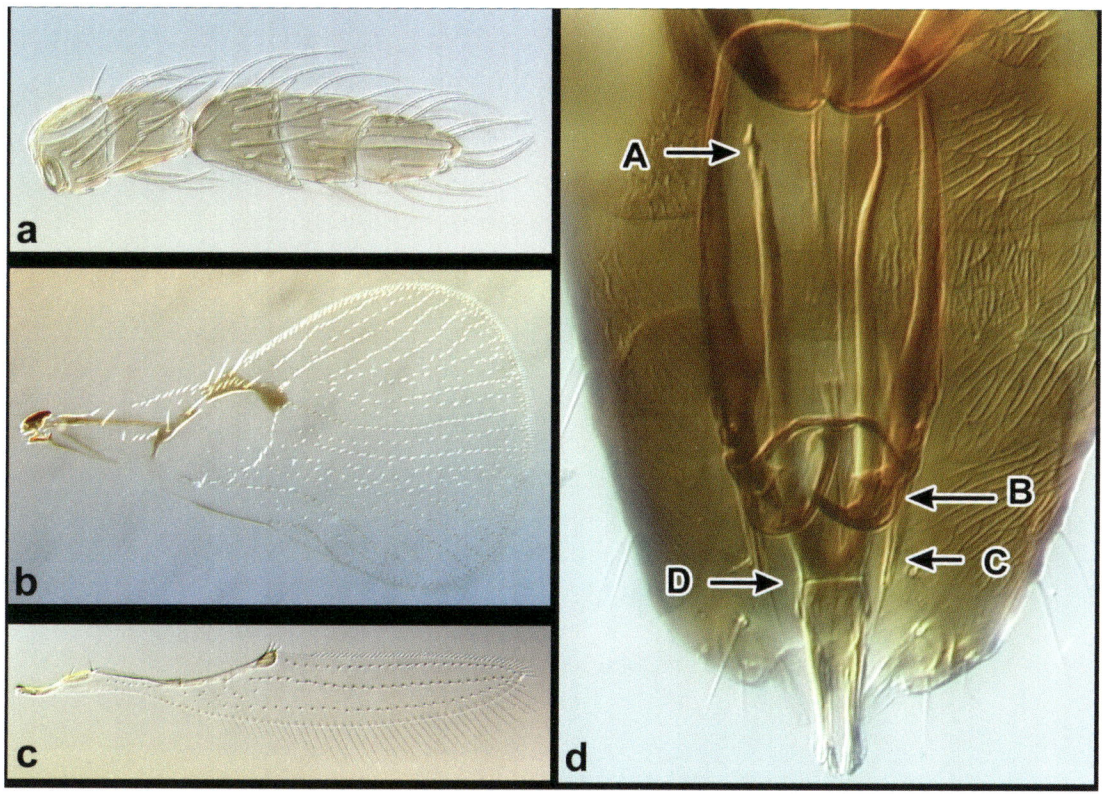

Figure 21. *Ufens decipiens*, ♂. (a) antenna, lateral; (b) forewing, dorsal; (c) hind
wing, dorsal; (d) genitalia, dorsal – arrows to {A} apodeme, {B} sickle-shaped
volsella, {C} paramere, {D} transverse hinge.

Ufens dilativena Nowicki, 1940, revised combination
(Fig. 22)
U. dilativena Nowicki, 1940: p. 625
Ufensia dilativena Viggiani 1988: p. 20 (new combination).

Diagnosis. - Forewing sparsely setose with narrowly diverging setal tracks r-m to M
and a single setal track between CU1 and CU2. Hind wing width not decreasing
immediately apical of hamuli. Mesoscutal sculpturing longitudinally striate.
Genitalia with parameres difficult to discern, short and straight, with a terminal
spine, their base even with posterior edge of anterodorsal aperture; no other
appendages present.

The genitalia of *U. dilativena* clearly ally it with species such as *U. forcipis*,
U. khamai, and *U. mezentius*. *U. dilativena* has parameres that are small and
difficult to discern but with a terminal spine, unlike the other species which have
large, prominent parameres without a terminal spine. Its parameres are also straight,

which separates it from *U. khamai* and *U. mezentius*. It is further separated from *U. khamai* by its lack of a ventral process.

Types. - ♦Lectotype, ♂ (DEZA), **here designated. CROATIA:** "Shumet, near Dubrovnik, 19.7.37" "coll. Nowicki". Paralectotype ♀, **BULGARIA:** "Varna, 1938" "coll. Nowicki". (DEZA).

Distribution. - Bulgaria, Croatia, Kenya, Madagascar, Namibia, South Africa.

Biology. - Known from host plants *Combretum* spp. and *Terminalia sericea* Burch. ex DC (Combretaceae). The host was listed as eggs of *Batchomorpha capeneri*, which is likely a misprint of *Batracomorphus capeneri* Linnavouri (Hemiptera: Cicadellidae).

Description. - BL 0.6 (0.5-0.6) mm; BL/HTL = 3.4 (3.1-3.8). Mesoscutal sculpturing longitudinally striate with interstitial sculpturing, when present, rugulose. Forewing sparsely setose, AA absent, single setal track between CU1 and CU2; FWL/HTL = 2.9 (2.8-3.0); FWL/FWW = 1.5 (1.4-1.6); FWFS/FWW = 0.07 (0.05-0.08); Max r-m to M/Min r-m to M = 1.8 (1.4-2.0); MV/PM = 0.9 (0.7-1.0); SV/MV = 1.3 (1.1-1.6); MV length/MV width = 2.1 (1.8-2.4). Hind wing width not immediately decreasing beyond hamuli; HWL/HWW = 7.5 (7.2-7.8); HWFS/HWW = 0.8 (0.8-0.9).

Male

Antenna: Club segments somewhat loosely joined, C/F = 2.9 (2.2-4.0); F2/F1 = 1.2 (1.1-1.3); APB absent on funicle; 1 PLS on each of F1-C2, 2 PLS on C3; 2-4 BPS on each of F1-C1, 1 BPS on C2 and C3; 8-11 FS on F1, 11-15 FS on F2, 10-15 FS on C1, 9-14 FS on C2, 8-11 FS on C3, 6-8 FS on C4; US absent on F1-C3.

Genitalia: GL/GW = 3.9 (3.3-4.8); GL/HTL = 1.1 (1.0-1.1); ADA/GL = 0.4; PAR difficult to discern, with terminal spine, subequal in width along entire length, their base posterior of posterior edge of ADA; PAR short, ca. 0.1 – 0.3 of GL; AI, AP, DR, VP, VS, transverse hinge absent.

Female

Antenna: C/F = 2.7 (1.9-3.9); F2/F1 = 1.8 (1.2-2.6); 1 APB on F1 and F2; 1 PLS on F1, 3-5 PLS on F2, 1-2 PLS on C1, 2 PLS on C2, 3-4 PLS on C3; 3-5 BPS on F1, 3-4 BPS on F2, 2-3 BPS on C1, 1 BPS on C2 and C3; 0 FS on F1 and F2, 4-6 FS on C1, 2-9 FS on C2, 1-2 FS on C3; 1 UPP on C3; 3-7 US on F1, 4 US on F2, 3-7 US on C1, 0 US on C2, 1-5 US on C3.

Ovipositor: OL/HTL = 1.2 (1.0-1.5).

Other Material Examined. **MADAGASCAR:** **Tulear:** Berenty, 12 km NW Amboasary, 5-15.v.1983, J. S. Noyes and M. C. Day, BM 1983-201 (2♂, 2♀). **Tamatave:** Perinet, 27.iv - 3.v.1983, J. S. Noyes and M. C. Day, BM 1983-201 (1 ♂, 1♀). **NAMIBIA:** Brandberg Wasserfallpläche, 21°13'0.5'S, 14°31'0.1'E, 1980m, 10-12.xi.1998, MT riverbed, A. H. Spriggs (1♂). **SOUTH AFRICA: Transvaal:** Warmbaths, ii.1964, A. L. Capener and D. P. Annecke, on *Combretum* spp. and *Terminalia sericea*, ex. eggs of *Batracomorpha* (as "*Batchomorpha*") *capeneri* Linnavouri (1♂, 1♀); Klaserie, 15 km NE (Guernsey Farm), 18-31.xii.1985, S. and J. Peck (1♂); **Gauteng:** Pretoria, ii.1957-iv.1961, D. Annecke, suction trap (many ♂ and ♀, on 20 slides).

Comments. - The only male and one of the two females described by Nowicki (1940) were examined. Of these syntypes, the male specimen is here designated as the lectotype. In addition to the above information, the lectotype male slide also has written on it "Ufens dilativena Nowicki ♂", "mat. tipico"[?], "pub G. Viggiani, 83". According to Nowicki (1940) the specimen was collected "along the railway at Shumet, near Dubrovnik, Dalmatia (estuary of Ombla River), 19 July, 1937." The female specimen is on a slide with the following information: "Ufens dilativena Nowicki ♀", "mat. tipico"[?], [Bulgaria] "Varna, 1938" "coll. Nowicki" "prep. G. Viggiani, 83". Nowicki (1940) reported a second female among the type series, which was reported from Yenina (Kazanlyk), Bulgaria. However, this specimen was not found. The female paralectotype is questionably conspecific. Although its wing characteristics are consistent with the lectotype, its collection in a different locality and without associated males hinders its confident placement.

Nowicki (1940) considered the most distinctive feature of *U. dilativena*, compared to species such as *U. foersteri* and *U. similis*, to be its thick marginal vein. *U. foersteri* and *U. similis* do indeed have a somewhat thin marginal vein, but *U. dilativena* is unremarkable among other *Ufens* species. Interestingly, the type series does seem to have slightly wider marginal vein than most of the conspecifics examined.

The disjunct geographical distribution between the type material and the specimens from southern Africa and Madagascar is somewhat perplexing. The male genitalia of the lectotype are somewhat difficult to see, though clearly very similar to those of the rest of the material examined. However, the parameres appear to be somewhat longer (ca. 0.3 of GL) and easier to discern compared to the African specimens (ca. 0.1 of GL). It is unknown if this is due to increased visibility due to the somewhat more lateral mount or if there is some fundamental difference between these genitalia, which may indicate different species. It seems likely that they comprise a single species as several specimens from Pretoria, South Africa, seem to have intermediate paramere length. Regardless, the difficulty of discerning the full extent of parameres prohibited measurement. The anteroventral edge of the genitalia is also difficult to discern in all individuals, making it difficult to accurately assess the anterior invagination.

The female specimens from South Africa contributed to all measurements except determination of antennal sensilla, as they were insufficiently cleared to confidently determine this parameter. Therefore, N=3 for all of these determinations. Interestingly, these South African females did seem to have comparatively much longer clubs (C/F = 3.7), than all other specimens (C/F = 2.1). Males from these respective localities do not exhibit any appreciable differences. Nonetheless, this species would benefit from further examination once further material becomes available.

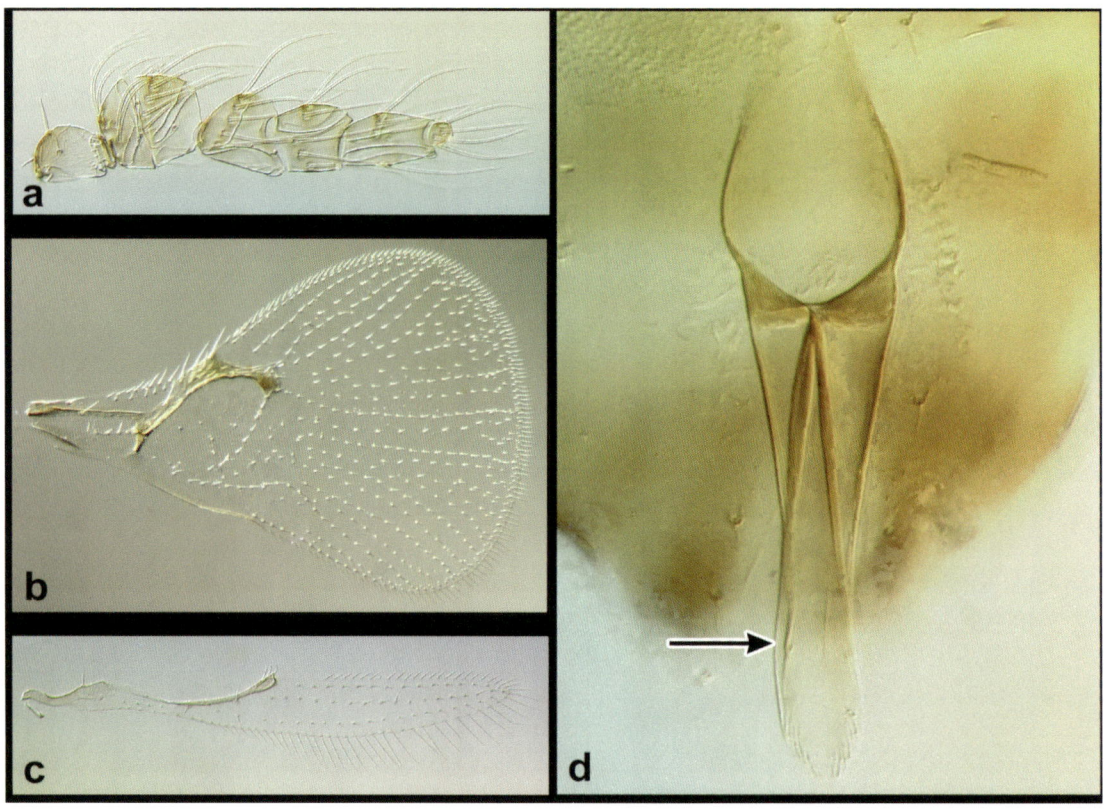

Figure 22. *Ufens dilativena*, ♂. (a) antenna, lateral; (b) forewing, dorsal; (c) hind wing, dorsal; (d) genitalia, dorsal – arrow to apex of paramere.

Ufens dolichopenis Owen, new species
(Fig. 23)

Diagnosis. - Forewing densely setose with widely diverging setal tracks r-m to M and dispersed setae between CU1 and CU2. Hind wing width decreasing immediately apical of hamuli. Mesoscutal sculpturing longitudinally cellulate (?). Genitalia capsule narrow, elongate, tubelike, gradually tapering to apex; length greater than (1.2-1.5x) hind tibial length; parameres short and straight, with a terminal spine, their base posterior to posterior edge of anterodorsal aperture, difficult to discern; volsellae likely present, but not distinguishable; no other appendages present.

  This species is clearly very closely allied to *U. simplipenis*, as the primary differentiating characteristics are genitalia and ovipositor lengths. These differences appear to be consistent within species and specimens have not been found with intermediate lengths. *U. dolichopenis* has male genitalia longer than the hind tibial length (1.2-1.5 x HTL), whereas *U. simplipenis* has genitalia shorter than hind tibial length (0.8-0.9 x HTL). Similarly, *U. dolichopenis* has an ovipositor longer than the hind tibial length (1.8-2.0 x HTL), whereas *U. simplipenis* has an ovipositor subequal to hind tibial length (0.9-1.0 x HTL).

Types. - ♦Holotype ♂, Allotype ♀ (USNM). **UNITED STATES: California:**
Kern Co.: Garlock, ca. 2 mi. SE, along Redrock-Randsburg Rd., 29.vii.1996, R.
Luck and JDP, d-Vac on *Petalonyx* sp. Paratypes 6♂, 4♀, same data (UCRC).

Etymology. - Conjunction of dolichos (Greek), long, and penis (Latin), in reference
to the elongate male genitalia.

Distribution. - México, United States.

Biology. - Unknown.

Description. - BL 0.7 (0.7-0.8) mm. BL/HTL = 3.4 (3.0-4.1). Mesoscutal
sculpturing longitudinally cellulate with interstitial sculpturing lightly rugulose.
Forewing densely setose; AA present; dispersed setae between CU1 and CU2;
FWL/HTL = 2.8 (2.4-3.2); FWL/FWW = 1.6 (1.3-1.7); FWFS/FWW = 0.06 (0.04-
0.08); Max r-m to M/Min r-m to M = 6.5 (5.4-9.3); MV/PM = 1.2 (1.2-1.3); SV/MV
= 0.8 (0.6-0.9); MV narrow, MV length/MV width = 3.6 (3.2-4.3). Hind wing width
decreasing immediately apical of hamuli; HWL/HWW = 9.6 (8.6-10.2);
HWFS/HWW = 1.2 (1.1-1.3).

Male

Antenna: Club segments somewhat loosely joined; C/F = 2.0 (1.8-2.0); F2/F1 = 1.0
(0.8-1.2); APB absent on funicle; 1 PLS on each of F1-C2, 2 PLS on C3; 2-5 BPS
on each of F1-C1, 1 BPS on C2 and C3; 10-15 FS on F1, 11-17 FS on F2, 9-15 FS
on C1, 10-18 FS on C2, 10-14 FS on C3, 8-12 FS on C4; US absent on each of F1-
C3.

Genitalia: Capsule long, thin, and gradually tapering; GL/GW = 5.6 (5.2-5.8);
GL/HTL = 1.3 (1.2-1.5); ADA/GL = 0.4 (0.4-0.5); AI slight to absent, AI/GL =
0.002 (0 -0.1); PAR minute, with terminal spine, subequal in width along entire
length, their base posterior to posterior edge of ADA; PAR/GL = 0.1 (0.08-0.1); VS
likely present, approximately subequal in length to PAR; transverse hinge, DR, AP,
VP absent.

Female (N=4)

Antenna: C/F = 2.2 (2.1-2.2); F2/F1 = 1.3 (1.0-1.5); 1 APB on F1 and F2, 0-1 APB
on C3; 1 PLS on F1, 2 PLS on each of F2-C2, 4 PLS on C3; 3-5 BPS on F1, 3-4
BPS on F2, 2-4 BPS on C1, 1 BPS on C2 and C3; 0 FS on F1 and F2, 6-8 FS on C1,
7-13 FS on C2, 3-8 FS on C3; 1 UPP on C3; 6-11 US on F1, 7-13 US on F2, 5-7 US
on C1, 0 US on C2, 1-2 US on C3.

Ovipositor: OL/HTL = 1.9 (1.8-2.0).

Other Material Examined. - **MÉXICO: Zacatecas:** Concepción del Oro, 4 mi.
NE, 4.vii.1984, JBW (1♂). **UNITED STATES: California:** *Los Angeles Co.*: San
Gabriel Mts., vic. Tie Cyn., 11.vi.1991, R. H. Crandell (1♂); *San Bernardino Co.*:
Holcomb Valley Rd. and Van Dusen Cyn. Rd., 16.vi.1988, R. K. Velten, SW
Ceanothus, etc. (4♂,1♀); Granite Mts. Reserve, Granite Cove, 34°48'N, 115°39'W,
14.v.1994, GP, SW (1♂); Lone Pine Cyn., 8 mi. SE Wrightwood, 14.v.1994, JDP,
SW (1♂).

Comments

In addition to the aforementioned morphological differences, molecular data also
suggest a partitioning of *U. dolichopenis* from *U. simplipenis*. While the practice of
using molecular divergences for species recognition is contentious, these two

species are separated by 7 bp in 28S-D2, a greater difference than that seen between other trichogrammatid species for which there are molecular data (AKO, unpublished). This molecular evidence, in addition to the male genitalic and ovipositor differences, provide confidence for recognizing *U. dolichopenis* as distinct. The collection records indicate that the two species may not be sympatric, as no single collection or collection site has yielded both species. Further research, both morphological and molecular, of this complex is recommended.

As in *U. simplipenis*, parameres in this species are very difficult to discern in slide-mounted specimens due to their small size and proximity to the capsule margin. PAR/GL measurement may therefore be prone to more error than other measurements. Volsellae in this species cannot actually be discerned in slide-mounted specimens, though they are likely present based on SEM micrographs of *U. simplipenis*. *U. dolichopenis* and *U. simplipenis* have the shortest parameres and volsellae relative to genitalia length known in the genus.

Molecular data for *U. dolichopenis* were presented in Owen et al. (2007) as *Ufens* sp. 7, and can be found under Genbank accession number AY623535 (28S-D2+D3).

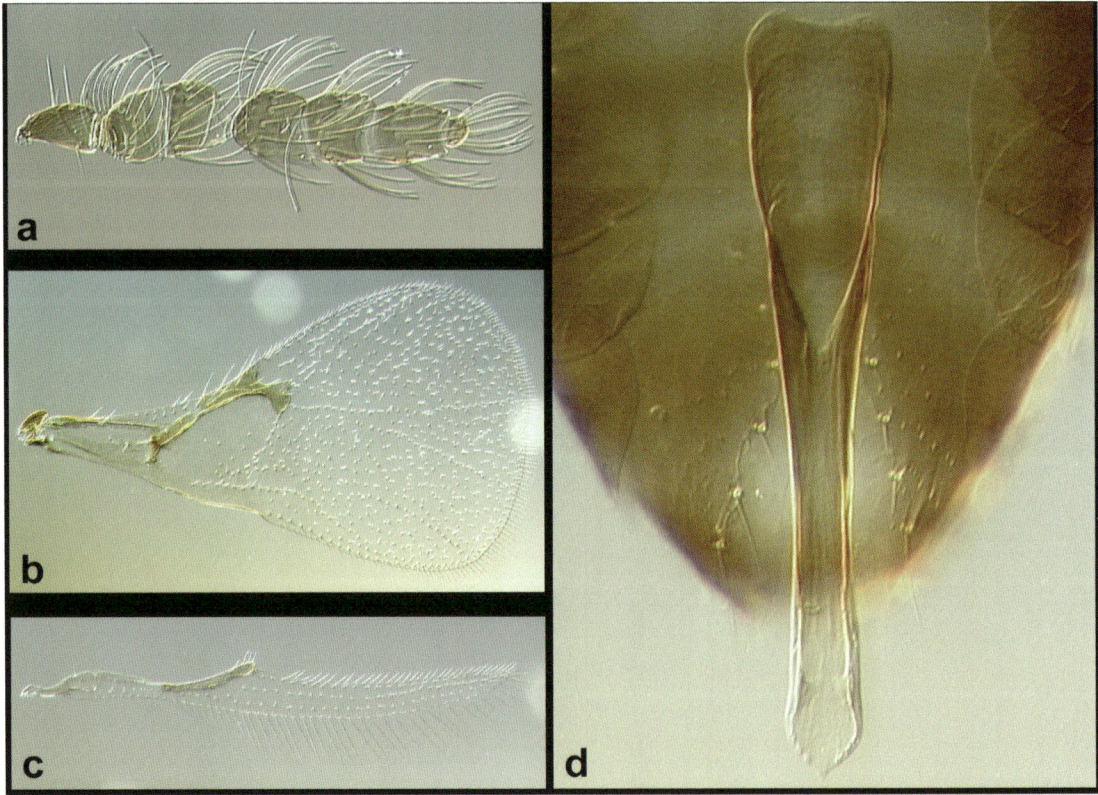

Figure 23. *Ufens dolichopenis*, ♂. (a) antenna, lateral; (b) forewing, dorsal; (c) hind wing, dorsal; (d) genitalia, dorsal.

Ufens elimaeae **Timberlake, 1927**
(Fig. 24)

U. elminaeae Timberlake, 1927: pp. 525-528.

Diagnosis. - Forewing sparsely setose with narrowly diverging setal tracks r-m to M and a single setal track between CU1 and CU2. Hind wing width decreasing slightly immediately apical of hamuli. Mesoscutal sculpturing longitudinally striate. Genitalia capsule narrow, obovate; anterodorsal aperture distinctly narrower than capsule; parameres widest at base, tapering, without a terminal spine, their base posterior to posterior edge of anterodorsal aperture; no other appendages present.

This species is perhaps most easily confused with *U. rimatus*, as both share similarly narrow genitalia in which the ADA is distinctly narrower than the maximum capsule width. *U. elimaeae* is separated by its stouter parameres without a terminal spine and by its lack of volsellae.

Types. - ◆Holotype ♀, Allotype ♂ (BPBM). **UNITED STATES**: **Hawaii**: *Oahu*: Tantalus, 16.i.1916, ex. eggs of *Elimaea punctifera* (Walker) [On a single slide with 1 other ♂ and 5 other ♀, designated as paratypes]. Other paratypes include: *Oahu*: Mt. Tantalus, ex. eggs of *Elimaea punctifera* in Koa leaf [*Acacia koa*?], 16.vi.1916, O. H. Sweezey [15♀, 2♂] (USNM); Waikiki, 16.xi.1919, R. W. Pemberton, ex. *Elimaea punctifera* [1♀ dissected] (BPBM); Barber's Point, 23.xii.1923, O. H. Swezey, ex. *Elimaea punctifera* [3♀] (BPBM).

Distribution. - Hawaiian Islands, including Hawaii, Kauai, Maui, Moloka'i, and Oahu.

Biology. - This species has been reared from *Elimaea punctifera* (Orthoptera: Tettigoniidae). Timberlake (1927) also lists *Holochlora japonica* Brunner (Orthoptera: Tettigoniidae) as a host, but the associated *Ufens* were not examined. This is the only *Ufens* species known from orthopteran hosts.

Description (N=4). - BL 0.6 (0.5-0.6) mm. BL/HTL = 3.6 (3.5-3.8). Mesoscutal sculpturing longitudinally striate with little interstitial sculpturing. Forewing sparsely setose; AA present; single setal track between CU1 and CU2; FWL/HTL = 3.1 (3.0-3.2); FWL/FWW = 1.5 (1.5-1.7); FWFS/FWW = 0.1 (0.09-0.1); Max r-m to M/Min r-m to M = 2.0 (1.7-2.3); MV/PM = 1.3 (1.2-1.5); SV/MV = 0.8 (0.8); MV length/MV width = 3.6 (3.3-4.2). Hind wing width decreasing slightly immediately apical of hamuli; HWL/HWW = 8.1 (8.0-8.3); HWFS/HWW = 1.2 (1.2-1.3).

Male

Antenna: Club segments compact; C/F = 2.0 (1.6-2.5); F2/F1 = 1.1 (1.0-1.2); APB absent on funicle; 1 PLS on each of F1-C2, 2 PLS on C3; 2-4 BPS on each of F1-C1, 1 BPS on C2 and C3; 4-6 FS on F1, 6-10 FS on F2, 6-9 FS on C1, 8-10 FS on C2, 5-8 FS on C3, 3-6 FS on C4; 0-1 US on F1, 2-4 US on F2, 0-3 US on C1.
Genitalia: Capsule narrow, obovate; GL/GW = 4.3 (4.3-4.4); GL/HTL = 1.3 (1.2-1.4); Maximum width of ADA distinctly narrower than capsule width; ADA/GL = 0.4 (0.4-0.5); AI shallow to absent, AI/GL = 0.006 (0-0.008); PAR width widest at base, without terminal spine, their base distinctly posterior to posterior edge of ADA; PAR/GL = 0.3; VP, AP, VS, DR, transverse hinge absent.

Female (N=3)

Antenna: C/F = 1.8 (1.6-1.9); F2/F1 = 1.4 (1.3-1.4); 1 APB on F1 and F2, 0-1 APB on C3; 1-2 PLS on F1, 3-4 PLS on F2, 1-2 PLS on C1, 2 PLS on C2, 4 PLS on C3; 4-5 BPS on each of F1-C1, 1 BPS on C2 and C3; 0 FS on F1 and F2, 5 FS on C1, 7-8 FS on C2, 2-3 FS on C3; 1 UPP on C3; 7-9 US on F1, 5-6 US on each of F2-C1, 0 US on C2, 1-3 US on C3.

Ovipositor: OL/HTL = 1.2 (1.2-1.3).

Other Material Examined. - **UNITED STATES**: **Hawaii**: *Hawaii*: Naulu Forest, Hawaii Volcanoes NP, 14.i.1971, J. W. Beardsley, ex. katydid eggs (1♂, 8♀) [1♂, 1♀ on 1st slide, 3♀ on 2nd slide, 4 ♀ on 3rd slide]; Hilo Coast, Koelekole Beach Park, 19.x.1983, D. M. LaSalle (1♀). *Kauai*: Opackaa Falls Lookout, 15.x.1983, D. M. LaSalle (2♂, 1♀). *Maui*: Lahaini, 23.xii.1928, D. H. Swezey (1♂). *Moloka'i*: Halawa Valley, 200' el., 29.ix-13.x.1995, W. D. Perreira, yellow sticky board trap (1♂); nr. Honomuni stream, 10' el., 2-16.ii.1996, W. D. Perreira, yellow sticky board trap (1♀); Mapuleh nr. Ililiopae Heiau, 10-40' el., W. D. Perreira, yellow sticky board trap (1♀). *Oahu*: Manoa Valley, 12-13.iii.1984, T. S. Bellows (1♂, 1♀); Manoa Valley, Lyon Arboretum, 19-21.v.1989, L. Masner, disturbed forest (1♂, 2♀; 1♀ card-mounted).

Comments. - The slide containing holotype and allotype specimens does not indicate the collector on the label, but Timberlake (1927) credits them to O. H. Swezey. This slide has the following notation "Holotype ♀ (best near center), allotype ♂ (larger) & paratypes." No other marks on the slide indicate which specimen is the holotype. However, the female specimen presumed to be the holotype is located near the center of the slide and has a more thoroughly cleared metasoma than other female specimens. Allotype determination is somewhat more problematic, though it is likely to be the specimen with its forewings and antenna more broadly overlapping, as it is marginally larger.

   U. elimaeae is the only *Ufens* known from the Hawaiian Islands; it is not known to occur elsewhere. It is therefore potentially an endemic species, though increased sampling in Oceania and Asia would be needed to confirm endemicity. The similarity of its genitalia to those of the Asian *U. rimatus* suggest historical ties with this region.

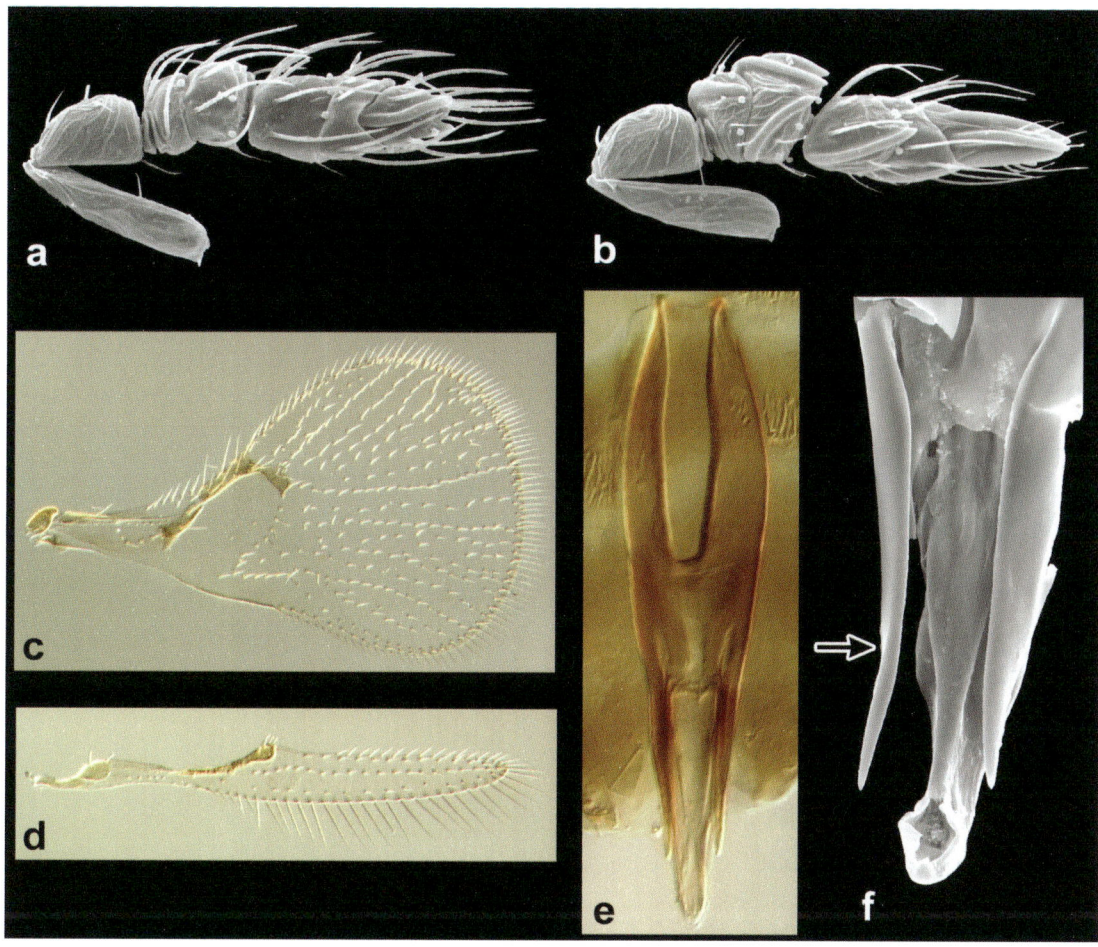

Figure 24. *Ufens elimaeae*. (a) ♂ antenna, lateral; (b) ♀ antenna, lateral; (c) forewing, dorsal; (d) hind wing, dorsal; (e) ♂ genitalia, dorsal; (f) ♂ genitalia, ventral – arrow to paramere.

Ufens flavipes **Girault, 1912**
(Fig. 25)

U. flavipes Girault, 1912: p. 72.
Dahms, 1984: pp. 609-610 (type material described)
Diagnosis. - Forewing sparsely setose with narrowly diverging setal tracks r-m to M and a single setal track between CU1 and CU2. Hind wing width not decreasing immediately apical of hamuli. Mesoscutal sculpturing longitudinally cellulate. Genitalia simplified; apodemes absent; ventral process elongate; transverse hinge present; no other appendages present.

As the only *Ufens* species lacking both parameres and volsellae, *U. flavipes* is unlikely to be confused with any other species. However, due to the simplified structure, its genitalia share a superficial resemblence to those of the North American *U. simplipenis* and *U. dolichopenis*. Unlike *U. flavipes* however, the latter species do not have a ventral process or a transverse hinge.

Types. - ◆Lectotype ♂ (QM), **here designated. AUSTRALIA: Queensland**: "Window of quarters, from Nelson [=Gordonvale], Q., 10.XII.1911, 3438, 778".

Distribution. - Australia.

Biology. - Unknown.

Description. - BL 0.6 (0.6-0.7) mm. BL/HTL = 3.6 (3.4-3.8). Mesoscutal sculpturing longitudinally cellulate with interstitial sculpturing rugulose. Forewing sparsely setose; AA present; single setal track between CU1 and CU2; FWL/HTL = 3.0 (2.9-3.1); FWL/FWW = 1.6 (1.5-1.7); FWFS/FWW = 0.06 (0.05-0.08); Max r-m to M/Min r-m to M = 2.4 (1.9-2.8); MV/PM = 1.0 (1.0-1.1); SV/MV = 1.0 (0.9-1.2); MV length/MV width = 2.4 (2.0-3.2). Hind wing width not decreasing immediately apical of hamuli; HWL/HWW = 7.6 (7.3-8.0); HWFS/HWW = 0.8.

Male

Antenna: Club segments compact. C/F = 2.1 (1.9-2.4); F2/F1 = 1.2 (1.0-1.5); APB absent on funicle; 1 PLS on each of F1-C2, 2 PLS on C3; 2-5 BPS on each of F1-C1, 1 BPS on C2 and C3; 9-12 FS on F1, 11-16 FS on F2, 11-13 FS on C1, 12-15 FS on C2, 6-11 FS on C3, 5-8 FS on C4; US absent on each of F1-C3.

Genitalia: Capsule widest at ca. half its length; GL/GW = 4.4 (4.0-5.2); GL/HTL = 1.0 (1.0-1.3); ADA/GL = 0.6 (0.5-0.6); AI shallow, AI/GL = 0.03 (0.02-0.04); VP wider at base, thin and tapering posteriorly, base < half of capsule width, VP/GL = 0.4; transverse hinge in apical third of capsule; DR present; PAR, AP, VS absent.

Female (N=2)

Antenna: Club compact, C/F = 1.7; F2/F1 = 1.6 (1.3-1.9); 1 APB on F1 and F2; 1 PLS on F1, 3 PLS on F2, 1-2 PLS on C1, 2 PLS on C2, 4 PLS on C3; 6 BPS on F1, 2-4 BPS on F2, 2-5 BPS on C1, 1-2 BPS on C2, 1 BPS on C3; 0 FS on F1 and F2, 4-7 FS on C1, 6-7 FS on C2, 1-2 FS on C3; 1 UPP on C3; 5-7 US on F1, 4-5 US on F2, 5-11 US on C1, 0 US on C2, 2-3 US on C3.

Ovipositor: OL/HTL = 1.4 (1.3-1.5).

Other Material Examined. - **AUSTRALIA: Northern Territory**: Darwin, 53 km SSW, 12°52'10.5"S, 130°35'04.4"E, 14-20.vii.1998, M. Hoskins, MT in mango patch (2♂); Arnhemland, 33 km E, Jabiru podocarp canyon, 15-23.xii.1993, S. and J. Peck (1♂). **Queensland**: Gordonvale, 22.xi.1979, E. C. Dahms, JBW, J. LaSalle (1♂); 9 km NW Mt. Tozer, 12°44'S 143°08'E, 30.vi-16.vii.1986, T. A. Weir, FIT in heath (1♂); 55.1 km W Rolleston, Hwy. 55, 13.iv.1988, JDP and G. Gordh, SW (1♂); Miles, 22.5 km N, 14.iv.1988, JDP and G. Gordh (1♂); 11 km NW Bald Hill, McIlwraith Ra., 500m. el., 26.vi-13.vii.1989, I. Naumann, MT in rainforest, search party campsite (2♂); Heathlands, 11°45'S 142°35'E, 15-26.i.1992, I. Naumann and T. Weir, MT (1♂); Heatlands, 11°45'S, 142°35E, 26.i-29.ii.1992, P. Feehney, MT (1♂); Cockatoo Crk. Xing, 17 km NW Heathlands, 11°39'S, 142°27'E, 22.iii-25.iv.1992, T. McLeod, MT open forest (2♂); Heathlands, 11°45'S, 142°35'E, 22.iii-25.iv.1992, T. McLeod, MT open forest (2♂); 7.1 km SE Chillagoe on Rd. to Mareeba, 17°12'21"S, 144°32'55"E, 2.iv.1992, E. C. Dahms and G. Sarnes (1♂); Heathlands dump, 11°45'S, 142°35'E, 25.vii-18.viii.1992, P. Zborowski and J. Cardale, MT open forest (1♂); Heathlands dump, 11°45'S, 142°35'E, 18.viii-17.ix.1992, P. Zborowski and L. Miller, MT open forest (2♂); Auburn River NP, Auburn Falls area, 151°04'E, 25°44'S, 23.ix.1995, JDP, SW 1° *Callistemon* (1♂);

Kin Kin, 9 km N, 25.ix.1995, JDP, SW forested area (2♂); Bribie Island (S end), 25.ix.1995, JDP, SW (1♂); Brisbane Forest Park, 27°25'04"S, 152°49'48"E, 5-12.xii.1997, N. Power, MT (1♂, 1♀); Brisbane Forest Park, 27°25'04"S, 152°49'48"E, 9-30.i.1998, N. Power MT (5♂); Brisbane Forest Park, 27°25'04"S, 152°49'48"E, 22-28.xi.1998, N. Power MT (1♂, 1♀); Mt. Isa, 12 km SW, 20°49'16"S, 139°27'38"E, 477m. el., 2.iii.2002, JMH, open *Eucalyptus* forest (1♂). **Western Australia**: 10 km N of Kununurra, Ivanhoe Crossing, 24.iii.1991, JDP, SW (1♂).

Comments. - Unlike most species described by Girault, a male is included in the type material and its genitalia are discernible. This male, designated as lectotype, is mounted on a slide with another specimen indicated as '*Oligosita minima*'. This *Ufens* is actually represented on 4 slides located in the Queensland Museum and 2 slides in the United States National Museum (Dahms 1984). All slides contain exclusively females, other than the slide containing the lectotype and another at the USNM. The male on the USNM slide is under a half coverslip and is designated as "*Ufens* sp." and is accompanied by a female under a complete coverslip designated as "*Ufens flavipes* Girault ♀". As the male is not given a specific epithet and it is not mentioned in the species description, it must be assumed that Girault considered this specimen a different species. It can now be recognized as *U. vectis* Owen, n. sp.

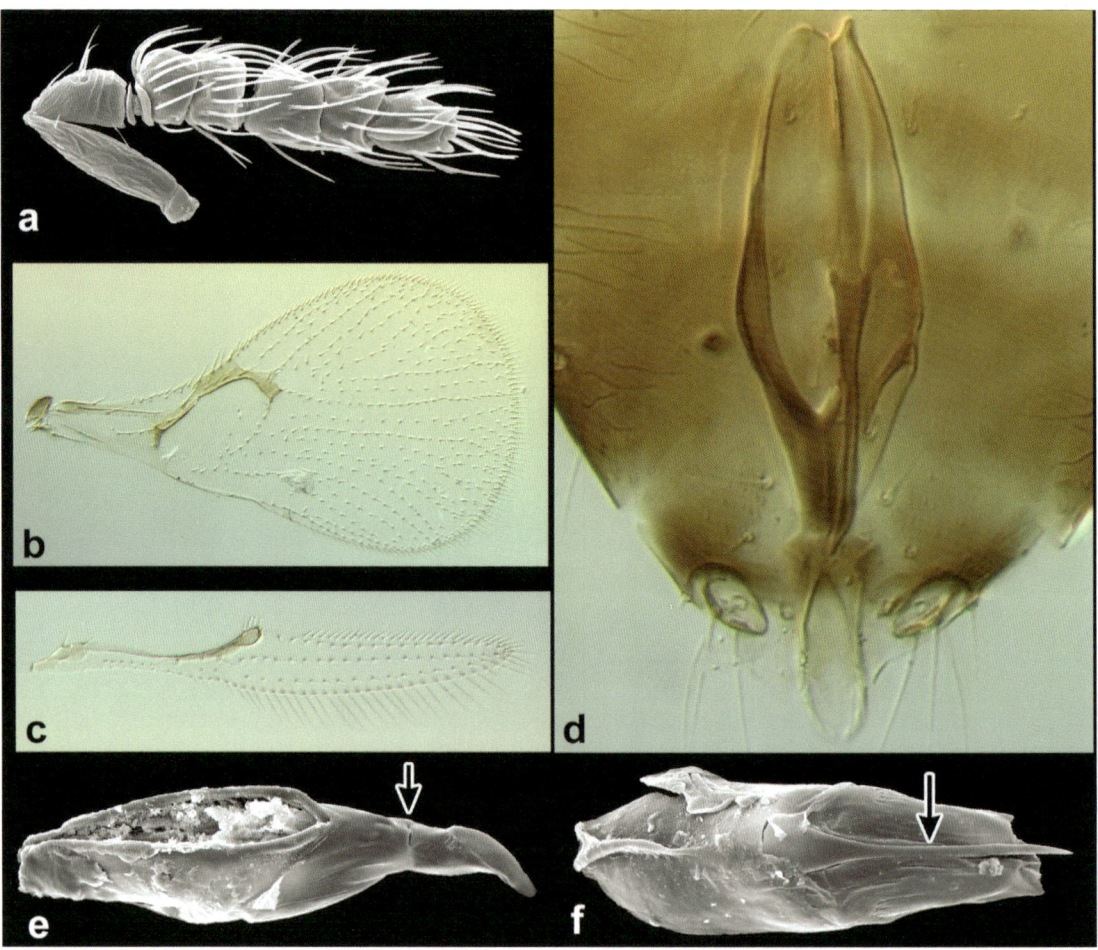

Figure 25. *Ufens flavipes*, ♂. (a) antenna, medial; (b) forewing, dorsal; (c) hind wing, dorsal; (d) genitalia, dorsal; (e) genitalia, dorsolateral – arrow to transverse hinge; (f) genitalia, ventral [apex broken off] – arrow to ventral process.

Ufens foersteri (Kryger, 1918), revised combination
(Fig. 26)

Centrobia foersteri Kryger, 1918: pp. 291-292.
Ufens hirticornis (Blood), 1923: p. 254.
Ufens foersteri irregularis Nowicki, 1935: p. 572.
Ufens foersteri meridionalis Nowicki, 1935: p. 572.
Viggiani, 1971: pp. 202-203 (*Ufensia* sp. genitalia illustrated [presumably *U. africana*]).
Ufensia africana Viggiani, 1972: pp. 159-161, **new synonymy.**
Ufensia minuta Viggiani, 1988: pp. 15-20, **new synonymy.**

Diagnosis. - Forewing sparsely setose with moderately diverging setal tracks r-m to M and a single setal track between CU1 and CU2. Hind wing width decreasing immediately apical of hamuli. Mesoscutal sculpturing longitudinally striate.

Genitalia laterally sigmoid; anterior invagination distinctly notch-like; parameres with terminal spine, subequal in width along entire length, their base anterior to posterior edge of anterodorsal aperture; no other appendages present.

Few other species are likely to be confused with *U. foersteri*. Although *U. simplipenis* shares its simplified genitalia and are also somewhat sigmoid laterally, *U. foersteri* has genitalia with a notch-like anterior invagation, and lacks volsellae. This species also has among the longest parameres when compared with genitalia length, whereas *U. simplipenis* has among the shortest.

Types. - ◆Holotype ♀ of *U. foersteri* (ZMUC). **DENMARK**: "2/8.1905. Fort. Indel., Mid [tugiens?], 16/7.1918. J. P. Kryger, Type." Wrapped in a piece of paper reading "*Ufens foersteri foersteri* (Kr.) 15-47".

Of *Ufensia africana* (new synonomy)

◆Holotype ♀, Allotype ♂ (DEZA) [single slide]. **GHANA**: "Ufensia africana Vigg., olotypo + allotypo, Ghana 10.v.66, C.I.E., D. S. Hill".

Of *Ufensia minuta* (new synonomy)

◆Holotype ♀ (DEZA). **ITALY**: "Ufensia minuta sp. n., ex. uova Reuteria marqueti, SU: Nocciolo, Domicella, 18.vi.85, Olotipo [in red], coll. prep. det. S. Viggiani".

Distribution. - Afrotropical, Australian, Indomalaysian, Palearctic.

Biology. - This species has been reared from *Reuteria marqueti* Puton (Hemiptera: Miridae) (Viggiani 1988) and possibly *Circulifer* sp. (Hemiptera: Cicadellidae). It has been collected in a broad range of habitats and reported from a number of different plants including: *Heliotropium* (Boraginaceae), *Chenopodium* sp. and *Salsola* sp. (Chenopodiaceae).

Description. - BL 0.7 (0.6-0.8) mm. BL/HTL = 4.1 (3.7-4.8). Mesoscutal sculpturing longitudinally striate with very little interstitial sculpturing. Forewing sparsely setose, AA present, single setal track between CU1 and CU2; FWL/HTL = 3.2 (3.0-3.5); FWL/FWW = 1.6 (1.5-1.7); FWFS/FWW = 0.08 (0.07-0.08); Max r-m to M/Min r-m to M = 2.9 (2.6-3.5); MV/PM = 1.1 (0.9-1.2); SV/MV = 1.0 (0.8-1.3); MV length/MV width = 2.9 (2.6-3.5). Hind wing width decreasing immediately apical of hamuli; HWL/HWW = 8.3 (7.6-8.7); HWFS/HWW = 1.0 (0.8-1.2).

Male

Antenna: C/F = 2.1 (1.9-2.4); F2/F1 = 1.2 (1.0-1.5); APB absent on funicle; 1 PLS on each of F1-C2, 2 PLS on C3; 2-4 BPS on each of F1-C1, 1 BPS on C2 and C3; 7-10 FS on F1, 9-12 FS on F2, 9-12 FS on C1, 8-10 FS on C2, 8-10 FS on C3, 6-8 FS on C4; US absent on each of F1-C3.

Genitalia: Capsule thin and laterally sigmoid; GL/GW = 7.5 (6-8.8); GL/HTL = 0.8 (0.7-0.9); ADA short, ADA/GL = 0.3; AI shallow but distinctly notch-like, AI/GL = 0.03 (0.02-0.05); PAR with terminal spine, subequal in width along entire length, their base distinctly anterior to posterior edge of ADA, long; PAR/GL = 0.6 (0.5-0.6); VP, AP, VS, DR, transverse hinge absent.

Female

Antenna: C/F = 2.0 (1.9-2.2); F2/F1 = 1.5 (1.4-1.6); 1 APB on F1 and F2, 0-1 APB on C3; 1 PLS on F1, 3 PLS on F2, 1-2 PLS on C1, 2 PLS on C2, 4 PLS on C3; 2-3 BPS on each of F1-C1, 1 BPS on C2 and C3; 0 FS on F1 and F2, 5-6 FS on C1, 5-9

FS on C2, 3-5 FS on C3; 1 UPP on C3; 6-9 US on each of F1-F2, 1-9 US on C1, 0-3 US on each of C2-C3.

Ovipositor: OL/HTL = 2.0 (1.3-2.6) (N=20).

Other Material Examined. - **AUSTRALIA: Queensland:** Great Sandy N.P., off Rainbow Beach Rd. (43), 26°00.62'S, 153°02.80'E, 16.xii.2002, JBM, AKO, SW grass/*Eucalyptus* forest (1♂). **Western Australia**: CALM site 28/3, 4km W of King Cascade, 15°38'S, 125°15'E, 12-16.vi.1988, T. A. Weir, MT closed forest (1♂). **CZECHOSLOVAKIA: Moravia**: Lanzhot-Ranspurk, 7-9.viii.1991, L. Masner, SW, climax flood forest (1♂); Lanzhot-Ranspurk, 7-9.viii.1991, L. Masner, YPT, climax flood forest (1♂); Lednice, 7-9.viii.1991, L. Masner, YPT, forest creek and pond (1♂); Palava nr. Mihulov, 9.viii.1991, L. Masner, SW (1♂). **ENGLAND**: Southampton: 'New Forest, Beaulieu Road, 18/VII.1921, 17-86', 'J. P. Kryger prep.' (1♂). **FRANCE:** Montpellier, 6-10.ix.1978, J. T. Huber (1♂); Montpellier, 23.vii.1979, J. T. Huber, SW at C.N.R.S. (1♂). **GERMANY: Ahmuhle**: nr. Hamburg, Sachsenwald, 22.viii.1984, L. Masner (1♂). **GHANA**: Tafo, 10.v.1966 (5♀, 2♂) [*U. africana* type series]. **GUINEA**: Mt. Nimba, Gouan River, 7°42'N, 8°23'W, 7-15.ii.1991, L. Leblanc, FIT rainforest; Mt. Nimba, Gouan River, 7°42'N, 8°23'W, 1-15.xii.1991, L. Leblanc, FIT rainforest. **IRAN: Tehran Prov.**: Karaj, 4.viii-3.ix.1977, J. T. Huber, YPT (3♂, 2♀). **ISRAEL**: Arava Valley, 0.2km N Hazeva Field School. 166m el., 30°46.77'N, 35°14.58'E, 9-31.iii.1995, M. E. Irwin, YPT and emergence trap (1♂). **ITALY: Basilicata**: Laghi di Monticchio, 650m el., 21.vi.1988, JDP (2♂, 1♀). **Campania**: *Avellino Prov.*: Domicella, 28.vi.1985, G. Viggiani, ex. eggs of *Reuteria marqueti* Puton (2♂, 6♀) [*U. minuta* type series]. *Benevento Prov.*: 1.8km E of Faicchio, 41°16.329'N, 14°29.884'E, 7.vi.2003, M. Bologna, JBM, AKO, JDP, SW (1♂, 1♀). **Lazio**: *Roma Prov.*: Caldara di Manziana, 42°05.61'N, 12°05.91'E, 305m el., 9-10.vi.2003, M. Bologna, JBM, AKO, JDP, YPT/SW *Quercus* forest edge, pasture (2♂, 2♀); Castelporziano Presidential Estate, La Focetta, 10m. el., 41°41.47'N, 12°22.63'E, 11.vi.2003, M. Bologna, JBM, AKO, JDP, SW riparian forest (1♂, 1♀); Castelporziano Presidential Estate, Ponte Guidoni, 41°45.415'N, 12°23.851'E, 11.vi.2003, M. Bologna, JBM, AKO, JDP, SW *Quercus ilex* forest (1♂, 2♀). *Viterbo Prov.*: San Giovenale, nr. Civitella Cesi, 225m el., 42°13.57'N, 12°00.04'E, 9.vi.2003, M. Bologna, JBM, AKO, JDP, SW Vesca creek, riparian (1♂); 5.5 km E Monte Romano, 42°15.28'N, 11°57.32'E, 305m el., 9-10.vi.2003, M. Bologna, JBM, AKO, JDP, YPT *Quercus* forest edge, pasture (1♂). **Puglia**: Otranto, 5 km S (Capo d'Otranto), 30.v.1992, JDP, SW flowers (1♀). **Sardinia**: Tempio, Cusseddu, 6-26.vi.1978 (2♂, 1♀). **Sicily**: Mandanici, 3km W, Monti Peloritani, 500m el., 3.vi.1992, JDP, SW (1♂, 1♀); Torri di Vendicari, 10 km N Pachino, 4.vi.1992, JDP, SW (1♀); ca. 20km S of Caltagirone (San Pietro area, hills to S), 5.vi.1992, JDP, SW mediterranean scrub (2♂, 1♀). **KENYA**: Kakamega District, Isecheno Nat. Res., 1800m. el., 0°14'24"N, 34°52'12"E, 21-28.ii.2002, R. Snelling, MT (1♂). **KYRGYSTAN: Dzhalal-Abad**: 18 km WSW Kazarman, 1550 m el., 41°22'1"N, 73°48'37"E, 15.vii.2000, C. H. Dietrich (2♂). **MADAGASCAR: Toliara Prov.**: Forêt de Mite, 20.7km WNW Tongobory, 75m. el., 23°31'27"S, 44°7'17"E, Fisher, Griswold et al., MT (1♂, 2♀). **MOROCCO**: Marrakech, Ouirgane, 1000m., 31°08'N, 08°05'W, 1996, C.

Kassebeer, MT (1♂). **NIGERIA:** Ibadan, IITA compound, x.1987, J. S. Noyes (1♂).
PAKISTAN: Punjab, Muree, 3km N, 5000' el., 24.v.1995, J. LaSalle, SW (3♂, 2♀).
RUSSIA: Moskow: Pushkino District, Mamontovka, 10-20.vii.2000, E Ya.
Shouvakhina, MT in garden (1♂). **SPAIN:** Aranquiez [location unidentifiable],
3.viii.1952, J. K. Holloway, on *Salsola* sp. (1♂); Cannelejos [location
unidentifiable], 22.vii-1.viii.1953, J. K. Holloway, ex.*Circulifer* sp. (?), em. in
Quarantine, Albany, California, USA (2♂); Barrajas [location unidentifiable],
8.viii.1953, J. K. Holloway, on *Chenopodium* , em. in Quarantine, Albany,
California, USA (1♂, 2♀); Barrajas [location unidentifiable], 14.viii.1953, J. K.
Holloway, on *Heliotropium*, em. in Quarantine, Albany, California, USA (1♂).
SOUTH AFRICA: East Transvaal: Kruger NP, Skukuza, 12-15.xii.1985, S. and J.
Peck (1♀); 15km NE Klaserie (Guernsey Farm), 18-31.xii.1985, S. and J. Peck, H.
and A. Howden (2♂, 1♀); 15km NE Klaserie (Guernsey Farm), 19-31.xii.1985, M.
Sanborne (3♂, 1♀). **TANZANIA:** Mkomazi Game Reserve, Ibaya Camp, 3.58°S,
37.48°E, 25.xii.1995-29.i.1996, S. van Noort, MT *Acacia, Commiphora,
Combretum* bushland (1♂). **THAILAND:** Surat Thani, Sok R., N of Hwy. 401,
8°54'26"N, 98°31'59"E, 20-21.ii.2005, D. Yanega (3♂). **TURKMENISTAN:** Old
Nisa, 9.vi.1992, S. Triapitsyn, on *Atriplex* sp. (1♂); Akhalskiy Etrap, Enev,
7.vi.1993, S. N. Myartseva, on *Atriplex* sp. (1♂). **UGANDA:** Laropi, viii.1941, T. H.
C. Taylor (2♂, 1♀).

Comments. - According to Kryger (1918), the holotype was captured in "Denmark:
Fortunens Indelukke" [an area in the west of Dyrehaven, north of Copenhagen] "in
Jaegersborg Dyrehave 5/8 1905. Swept in short dry grass on main road."

This species has the widest known distribution of any *Ufens* species (Table
3) as it is known from Aftropical, Australian, Indomalaysian, and Palearctic regions.
In addition to some of the above countries, Lin (1994) also lists it as present in
Egypt, Greece, Poland, Romania, and Turkey. This is a somewhat perplexing
species, especially in light of its broad geographic range. Males from different
geographical regions cannot be differentiated from each other based on forewing or
genitalic characters. However, the females that are likely associated with these
males do show widely diverging ovipositor lengths (OL/HTL varying from 1.3-2.6).
There does not appear to be much variation among females from an individual
collection. For example in four paratypes of *U. africana* (herein synonomized) the
OL/HTL range is only 2.2 – 2.6. However, there does not appear to be a geographic
cline in ovipositor lengths, with specimens from Italy spanning nearly the entire
observed range (OL/HTL varying from 1.6-2.5), though specimens from most other
localities demonstrate less variation. For comparison, the nominal female holotype
possesses an OL/HTL of 2.3. It seems possible that what is herein conceived as a
single species may prove to be a closely related group of species, though further
information is needed. For now, the continuous variation of this character argues
against this possibility.

Viggiani (1988) separated *U. dilativena* from *U. foersteri* and *U. minuta*
(herein synonomized) by the relative width of the marginal vein and forewing. The
width of the forewing was found in this study to overlap in these nominal species.
However, the length/width of the marginal vein does not overlap as *U. foersteri* has

a thinner marginal vein (similar to *U. similis*) than *U. dilativena*. Marginal vein width does not appear to be diagnostic of specimens identified as either *U. foersteri* or *U. minuta*, as their ranges overlap. Viggiani (1988) further separated *U. foersteri* and *U. minuta* by the length of the ovipositor and relative width of the marginal vein. As indicated, there is considerable but continuous variation in ovipositor length in this group. Regarding *U. africana*, Viggiani (1988) did not address differences between this species and *U. minuta* or *U. foersteri*. No discrete differences were found between specimens identified as these species and their male genitalia are identical.

The definition of *U. foersteri* was based on females; males were unknown to Nowicki. This prompts the question of whether the primarily male-based definition of *U. foersteri* adopted here is consistent with the original description. The limited *Ufens* diversity in Western Europe allows considerable confidence that the species is correctly identified. Firstly, the only other species known from the region are *U. dilativena*, known in Europe only from its types, and the widespread *U. similis*. Females of both species can be separated from those of *U. foersteri* – the forewing of *U. similis* is considerably more setose; the hindwing of *U. dilativena* does not decrease in width beyond the hamuli and it has a thicker marginal vein. Secondly, females consistent with the holotype of *U. foersteri* have been collected in Europe with males having the same structure of non-sexually dimorphic features. The only possibility of error is that the female holotype of *U. foersteri* is associated with males that have not been previously collected. Further collections in the type locality may be able to alleviate this final possibility of error.

Type specimens of *U. foersteri irregularis* and *U. foersteri meridionalis* were not located for this study. The synonymy of these taxa indicated by Doutt and Viggiani (1968) is presumed correct. Similarly, the type specimens of *U. hirticornis* (originally described as *Neocentrobia hirticornis*) were not located. In the above cases, synonomy with *U. foersteri* is inferred based upon all known information, but should be reevaluated when the type specimens are found.

Molecular data for *U. foersteri* were presented in Owen et al. (2007) as *Ufensia minuta*, and can be found under Genbank accession numbers AY623541 (28S-D2+D3) and AY940382 (18S).

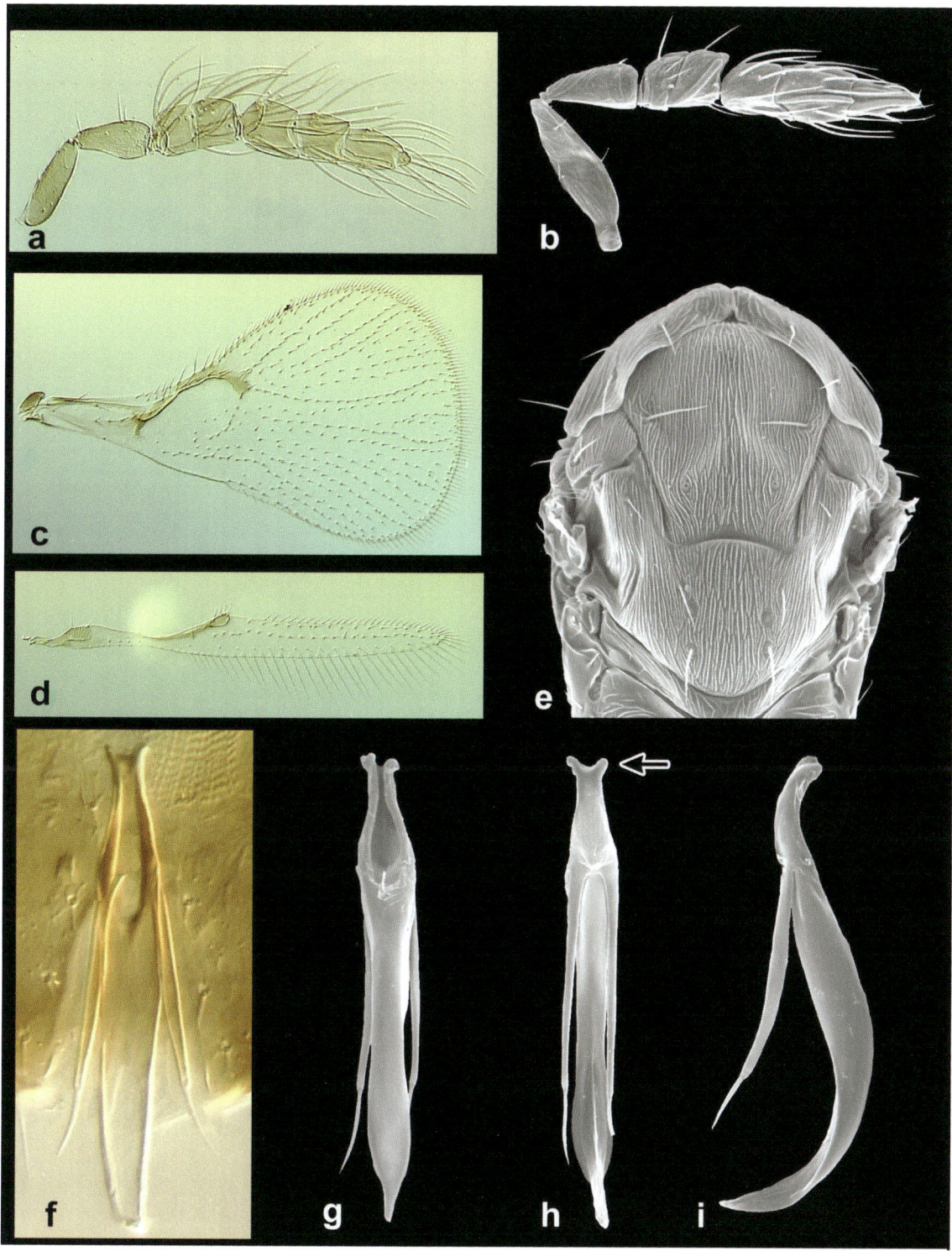

Figure 26. *Ufens foersteri*. (a) ♂ antenna, lateral; (b) ♀ antenna, lateral; (c) forewing, dorsal; (d) hind wing, dorsal; (e) mesosoma, dorsal; (f-g) ♂ genitalia, dorsal; (h) ♂ genitalia, ventral – arrow to notch-like anterior invagination; (i) ♂ genitalia, lateral.

Ufens forcipis Owen, new species
(Fig. 27)

Diagnosis. - Forewing sparsely setose with narrowly diverging setal tracks r-m to M; a single setal track between CU1 and CU2. Hind wing width not decreasing immediately apical of hamuli. Mesoscutal sculpturing longitudinally striate. Genitalia with anterodorsal aperture broad, nearly transversely oval; parameres long, nearly straight, tapering, without a terminal spine, their base even with posterior edge of anterodorsal aperture; no other appendages present.

The simplified genitalia with only long parameres and lacking a terminal spine suggest a relationship with *U. khamai* and *U. mezentius*. However the unique, nearly transversely oval shape of the anterodorsal aperture, combined with nearly straight parameres will separate this species from both.

Types. - ◆Holotype ♂ (CNC). **OMAN**: Muscat, Madinat Qaboos, 20-28.ii.1986, pans and SW, J.T. Huber (CNC). Paratype ♂, same data (UCRC).

Etymology. - Latin for forceps, in reference to the long, straight parameres.

Distribution. - Oman.

Biology. - Unknown.

Description (N=2). - BL 0.5 (0.5) mm. BL/HTL = 3.4. Mesoscutal sculpturing longitudinally striate without obvious interstitial sculpturing. Forewing sparsely setose; AA absent; single setal track between CU1 and CU2; FWL/HTL = 3.2 (3.1-3.3); FWL/FWW = 1.6; FWFS/FWW = 0.1; Max r-m to M/Min r-m to M = 1.8 (1.7-1.8); MV/PM = 1.0 (0.9-1.2); SV/MV = 0.9 (0.8-1.0); MV length/MV width = 2.1 (1.9-2.4). Hind wing width not decreasing immediately apical of hamuli; HWL/HWW = 7.6 (7.3-7.9); HWFS/HWW = 1.1 (1.0-1.1).

Male

Antenna: Club segments compact; C/F = 2.0; F2/F1 = 1.1 (1.1-1.2); APB absent on funicle; 1 PLS on each of F1-C2, 2 PLS on C3; 1-2 BPS on each of F1-C1, 1 BPS on C2 and C3; 9-11 FS on F1, 12-16 FS on F2, 12-14 FS on C1, 12-15 FS on C2, 10-11 FS on C3, 7-8 FS on C4; 0 US on F1, 0-1 US on F2, 0-2 US on C1.

Genitalia: Capsule equally wide through ADA, then narrow but parallel through most of the remainder of its length; GL/GW = 1.4 (1.3-15); GL/HTL = 0.7; ADA nearly transversely oval, ADA/GL = 0.4 (0.3-0.4); AI shallow, AI/GL = 0.07 (0.05-0.09); PAR without terminal spine, straight for most of its length and only slightly diverging from midline apically, width greatest near base and slowly tapering, their base even with posterior edge of ADA; PAR/GL = 0.7 (0.6-0.7), PAR/PAR width = 7.3 (7.2-7.5); transverse hinge, AP, DR, VP, and VS absent.

Female
Unknown.

Other Material Examined. - None.

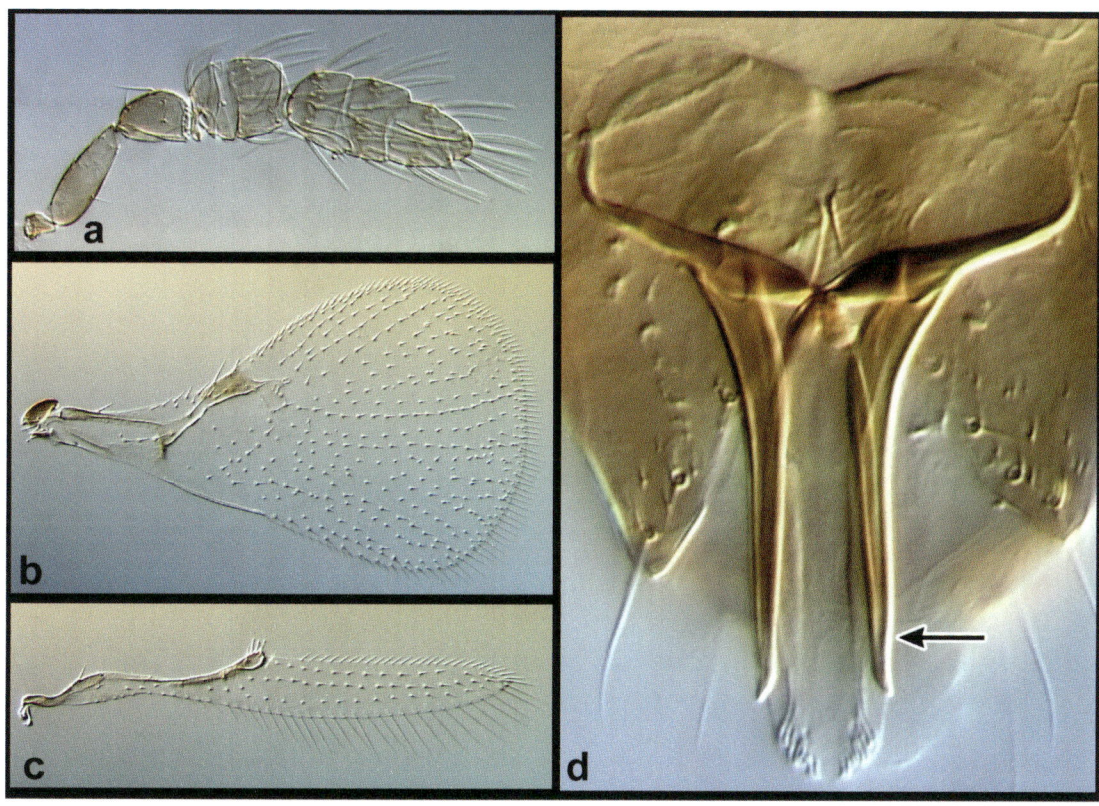

Figure 27. *Ufens forcipis*, ♂. (a) antenna, lateral; (b) forewing, dorsal; (c) hind wing, dorsal; (d) genitalia, dorsal – arrow to paramere.

Ufens gloriosus Owen, new species
(Fig. 28)

Diagnosis. - Antenna with aporous sensillar trichodea B on funicle; numerous placoid sensilla on all funicle and club segments except C4; flagelliform setae absent on first funicle segment but abundant on remaining funicle and club segments. Forewing sparsely setose with narrowly diverging setal tracks r-m to M; a single setal track between CU1 and CU2. Hind wing width increasing apical of hamuli. Mesoscutal sculpturing longitudinally cellulate. Genitalia with apodemes present; parameres with a terminal spine, their base anterior to posterior edge of anterodorsal aperture; volsellae somewhat sickle-shaped with ends curved in towards midline; ventral process present.

This species has very distinctive genitalia, and is unlikely to be confused with any other species except *U. decipiens*. *Ufens gloriosus* can be differentiated, however, by the presence of alar acanthae, lack of unsocketed setae on the funicle, greater number of placoid sensilla on the funicle and club segments, and lack of flagelliform setae on F1. Considering their unique genitalia within *Ufens*, it is possible that *U. decipiens* and *U. gloriosus* represent a single species with polymorphic antenna. However, such diversity has not been seen in other species, even those represented by much larger series of individuals. In addition, no quantitative variation was observed in antenna sensillar distribution between *U. decipiens* and *U. gloriosus*, lending additional support to the notion that they are separate.

Types. - ♦Holotype ♂ (QM). **AUSTRALIA**: **Queensland**: 6.1km SE Chillagoe on Rd. to Mareeba, 17°09'30"S 144°31'25"E, 26.iii.1992, E. C. Dahms and G. Sarnes.

Etymology. - Latin for glorious.

Distribution. - Australia.

Biology. - Unknown.

Description (N=3). BL 0.9 (0.8-1.0) mm. BL/HTL = 3.7 (3.3-4.1). Mesoscutal sculpturing longitudinally cellulate with interstitial sculpturing primarily transverse. Forewing sparsely setose, AA absent, single setal track between CU1 and CU2; FWL/HTL = 2.9 (2.7-3.0); FWL/FWW = 1.6 (1.5-1.6); FWFS/FWW = 0.03 (0.02-0.03); Max r-m to M/Min r-m to M = 1.8 (1.6-2.0); MV/PM = 0.9 (0.8-0.9); SV/MV = 1.3 (1.1-1.4); MV length/MV width = 2.3 (2.0-2.8). Hind wing width increasing apical of hamuli; HWL/HWW = 6.3 (6.0-6.7); HWFS/HWW = 0.6 (0.5-0.9).

Male

Antenna: Club segments very compact; C/F = 2.0 (1.9-2.0); F2/F1 = 1.4 (1.2-1.5); 2 APB on F1, 1 APB on F2; 6-9 PLS on F1, 6-8 PLS on F2, 4-6 PLS on C1, 6-9 PLS on C2, 4-7 PLS on C3; 2-7 BPS on each of F1-C1, 1 BPS on C2 and C3; 0 FS on F1, 8-12 FS on F2, 14-24 FS on C1, 22-25 FS on C2, 12-16 FS on C3, 1-3 FS on C4; 0 US on each of F1-F2, 1 US on C1.

Genitalia: Capsule nearly parallel sided in anterior half, anterior margin nearly transverse; GL/GW = 3.3 (3.1-3.4); GL/HTL = 1.2 (1.1-1.2); ADA/GL = 0.6; AI extremely shallow, AI/GL = 0.01; PAR with terminal spine, subequal in width along entire length, their base distinctly anterior to posterior edge of ADA; PAR/GL = 0.3; VS unusual, somewhat sickle-shaped with ends curved in towards midline; AP length/GL = 0.5 (0.4-0.6); transverse hinge in apical quarter of genitalia; DR present, extending beyond quarter of GL; VP base <half maximum width of genital capsule, small, VP/GL = 0.2 (0.2-0.3).

Female

Unknown.

Other Material Examined. - **AUSTRALIA**: **Australian Capital Territory**: Canberra, Black Mtn., 8-13.iii.1999, G. Gibson, YPT (1♂) (QM). **South Australia**: Orapinna Crk., Dingly Dell Camp nr. water, 31°21'S, 138°42'E, 4-10.xi.1990, I. Naumann and J. Cardale, MT (1♂) (QM).

Comments. - *U. gloriosus* has been collected together with *U. decipiens* in the Australian Capital Territory, the only time either species has been collected in that region. The fact that their minor character differences hold in sympatry provides additional evidence for species recognition.

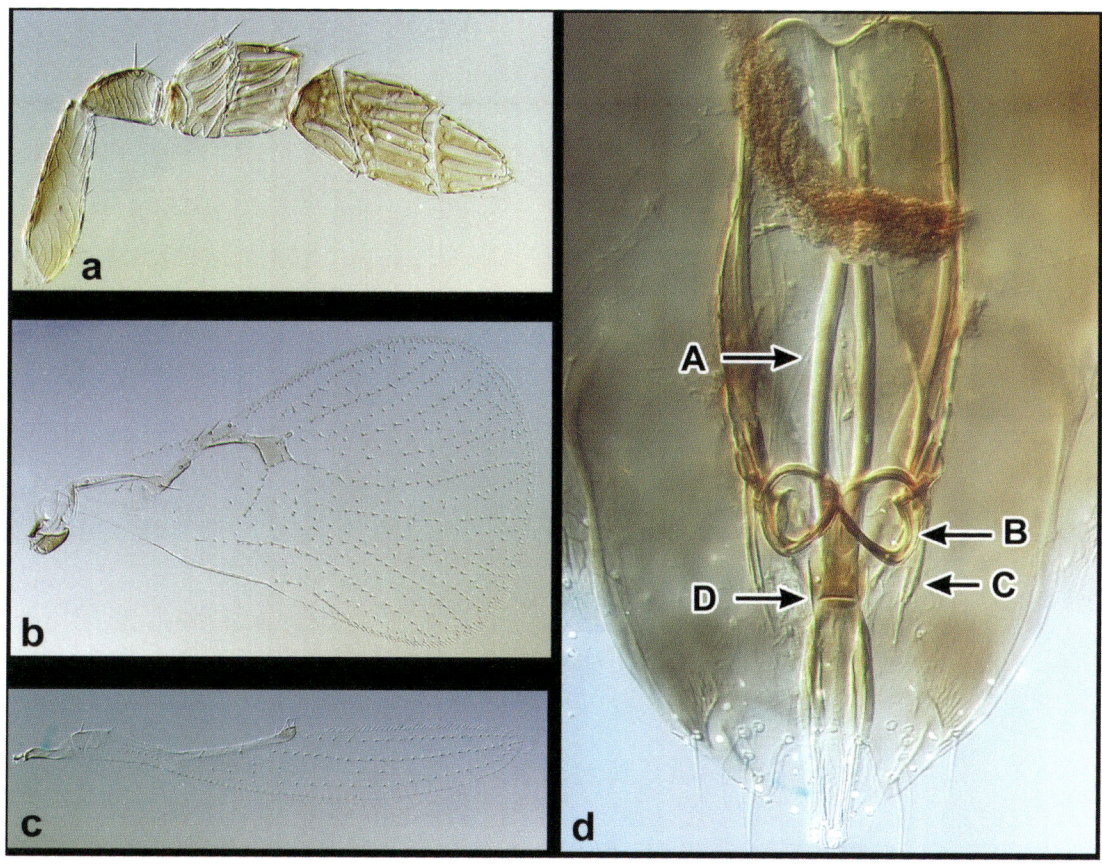

Figure 28. *Ufens gloriosus*, ♂. (a) antenna, lateral; (b) forewing, dorsal; (c) hind wing, dorsal; (d) genitalia, dorsal – arrows to {A} apodeme, {B} sickle-shaped volsella, {C} parameres, {D} transverse hinge.

Ufens hercules **Girault, 1912**
(Fig. 29)

U. hercules Girault, 1912: p.73.
Dahms, 1984: p. 699 (type material described)

Diagnosis. - Forewing sparsely setose with narrowly diverging setal tracks r-m to M; a single setal track between CU1 and CU2. Hind wing width increasing apical of hamuli. Mesoscutal sculpturing longitudinally striate. Genitalia with marked anterior invagination; parameres with a terminal spine, their base posterior to posterior edge of anterodorsal aperture; volsellae straight and, together with parameres, defining lateral edge of capsule; paired terminal spines near apex of genitalia.

The small spines near the apex of the genitalia are unique among *Ufens*, though they are somewhat remiscient of the terminal spines found on the dorsal apex of the genitalia of some other trichogrammatids, such as *Nicolavespa* (Pinto 2006). However, they appear to be located on the ventral surface, and not on the dorsal surface as found in *Nicolavespa*. The spatulate shape of *U. hercules* genitalia is similar to that found in species such as *U. gloriosus* and *U. cardalia*, but its lack of apodemes and ventral process immediately distinguish it.

Types. - ◆Holotype ♂ (QM). **AUSTRALIA: Queensland:** "Queensland Museum. 3440. Type, Hy/779, ♂", "*Ufens hercules* Girault, Type. *Aphelinoidea howardii, huxleyi* Girault. From window of a carhouse, Mareeba, N. Q., Jany 2.1912, AAG. 779".

Distribution. - Australia.

Biology. - Unknown.

Description. - BL 0.6 (0.6-0.7) mm. BL/HTL = 3.5 (3.4-3.6). Mesoscutal sculpturing longitudinally striate with interstitial sculpturing rugulose. Forewing sparsely setose, AA absent, single setal track between CU1 and CU2; FWL/HTL = 2.8 (2.7-2.8); FWL/FWW = 1.5 (1.5-1.6); FWFS/FWW = 0.08 (0.07-0.08); Max r-m to M/Min r-m to M = 1.7 (1.5-2.1); MV/PM = 1.1 (1.0-1.1); SV/MV = 1.0 (0.9-1.2); MV length/MV width = 3.3 (2.8-4.1). Hind wing width increasing apical of hamuli; HWL/HWW = 6.6 (6.2-7.1); HWFS/HWW = 0.7 (0.07-0.08).

Male

Antenna: C/F = 2.3 (2.2-2.4); F2/F1 = 1.2 (0.9-1.4); APB absent on funicle; 1 PLS on each of F1-C2, 2 PLS on C3; 1-5 BPS on each of F1-C1, 1 BPS on C2 and C3; 6-10 FS on F1, 8-14 FS on F2, 8-13 FS on C1, 8-14 FS on C2, 7-10 FS on C3, 5-6 FS on C4; 0 US on each of F1-C3.

Genitalia: Capsule long, spatulate to obovate; GL/GW = 4.1 (3.9-4.3); GL/HTL = 1.5 (1.4-1.6); ADA/GL = 0.5 (0.4-0.5); AI pronounced, AI/GL = 0.1 (0.09-0.1); PAR with terminal spine, subequal in width along entire length, their base distinctly posterior to posterior edge of ADA; PAR/GL = 0.2 (0.2-0.3); VS straight and rigid, base near midline, and apparently with medial spine at ca. 2/3 GL, VS/GL = 0.5 (0.5-0.6); apex of genital capsule with well-sclerotized pair of small spines on ventrum; transverse hinge present, though difficult to discern in some specimens; AP, DR, VP absent.

Female

Unknown.

Other Material Examined. - **AUSTRALIA: Queensland:** Cockatoo Creek Crossing, 17 km NW Heathlands, 11.39°S, 142.28°E, 26.i-29.ii.1992, P. Feehney, MT open forest (4♂); Cockatoo Creek Crossing, 17 km NW Heathlands, 11.39°S, 142.28°E, 22.iii-25.iv.1992, T. McLeod, MT open forest (5♂); Heathlands, 11.45°S, 142.36°E, 26.i-21.iii.1992, P. Feehney, MT (3♂); Heathlands, 11.45°S, 142.36°E, 22.iii – 7.vi.1992, T. McLeod, MT open forest (3♂); Heathlands dump, 11.45°S, 142.36°E, 25.vii-18.viii.1992, P. Zborowski and J. Cardale, MT open forest (1♂); Heathlands dump, 11.45°S, 142.36°E, 18.viii-17.ix.1992, P. Zborowski and L. Miller, MT open forest (1♂); Heathlands dump, 11.45°S, 142.36°E, 20.x-21.xii.1992, P. Zborowski and A. Calder, MT open forest (1♂).

Comments. - The holotype is mounted under a half cover slip on a slide with two female *Aphelinoidea*, and is lateral in position with the genitalia somewhat extruded, and wings and antenna in fairly good condition. As *U. hercules* was described from a single male with visible genitalia, its identifiability is more straighforward than most other species described by Girault.

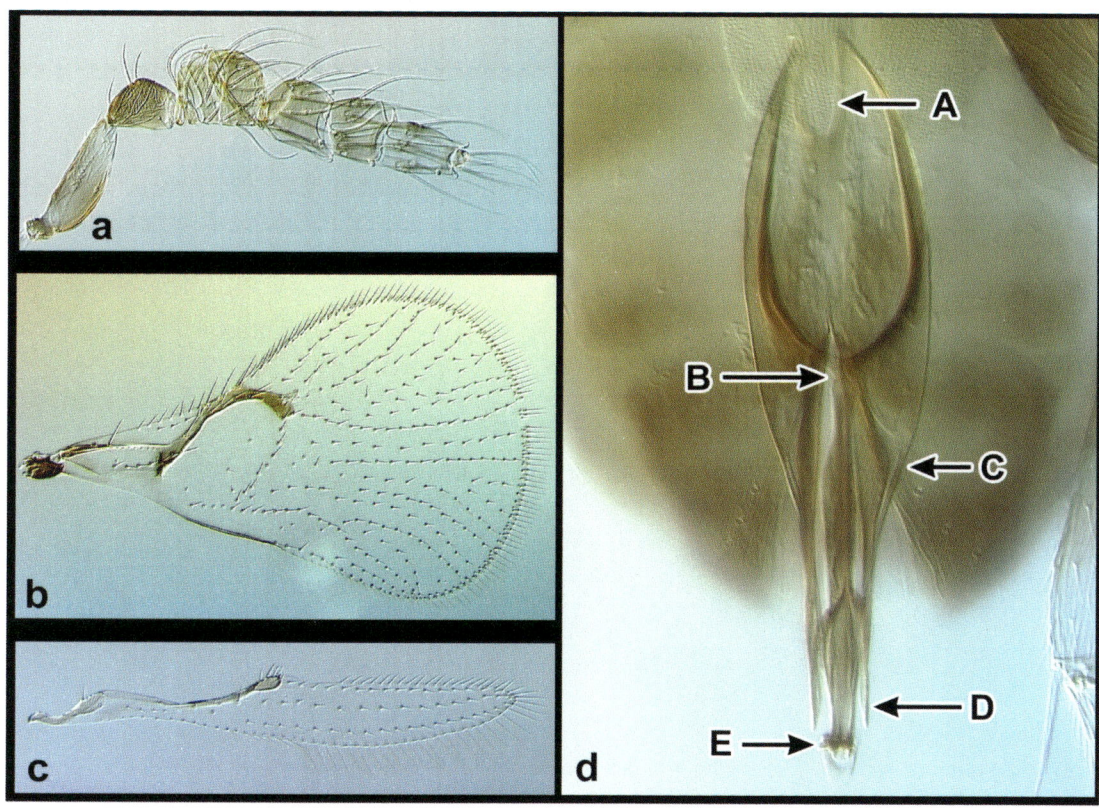

Figure 29. *Ufens hercules*, ♂. (a) antenna, lateral; (b) forewing, dorsal; (c) hind wing, dorsal; (d) genitalia, dorsal – arrow to {A} anterior invagination, {B} transverse hinge, {C} paramere base, {D} volsella apex, {E} terminal spines.

Ufens invaginatus Owen, new species
(Fig. 30)

Diagnosis. - Forewing sparsely setose with narrowly diverging setal tracks r-m to M; a single setal track between CU1 and CU2. Hind wing width decreasing apical of hamuli. Mesoscutal sculpturing longitudinally striate. Genitalia with very pronounced anterior invagination; apodemes absent; anterodorsal aperture distinctly narrower than width of capsule; parameres with a terminal spine, their base even with posterior edge of anterodorsal aperture, and apically diverging from midline; volsellae straight and rigid, immediately adjacent to apex of genital capsule for most of their length.

The depth of the anterior invagination is unique among *Ufens* species. The volsellae are somewhat difficult to distinguish in slide-mounted specimens as they coincide with the lateral edges of the capsule. However, in several specimens the apical portion of the volsellae does diverge, showing that they are separate structures and not simply thickened lateral edges of the capsule. *U. invaginatus* is one of the few species in which the anterodorsal aperture is distinctly narrower than the maximum width of the capsule, though the deep anterior invagination easily separates it from the other species with this trait. The only other species known to have a similarly pronounced anterior invagination is *U. ceratus*. However, *U. invaginatus* is readily distinguished by the presence of a ventral process and parameres which are widest near the middle.

Types. - ◆Holotype ♂ (QM). **AUSTRALIA: Queensland:** Bribie Island (S end), 25.ix.1995, JDP, SW (QM). Paratypes 4♂, 1♀ same data (1♂, 1♀ QM; 1♂ UCRC)

Etymology. - Latin for to fold or draw back within itself, in reference to the deeply invaginated male genitalia of this species.

Distribution. - Australia.

Biology. - Unknown.

Description. - BL 0.6 (0.5-0.6) mm. BL/HTL = 3.0 (2.6-3.3). Mesoscutal sculpturing longitudinally striate with interstitial sculpturing lightly rugulose to longitudinal. Forewing sparsely setose, AA present, single setal track between CU1 and CU2; FWL/HTL = 3.0; FWL/FWW = 1.7 (1.6-1.8); FWFS/FWW = 0.08 (0.07-0.09); Max r-m to M/Min r-m to M = 2.0 (1.8-2.2); MV/PM = 1.2 (1.0-1.5); SV/MV = 0.9 (0.8-1.3); MV length/MV width = 3.2 (2.7-4.2). Hind wing width decreasing immediately apical of hamuli; HWL/HWW = 8.8 (8.5-9.0); HWFS/HWW = 1.1 (1.0-1.1).

Male

Antenna: C/F = 2.2 (2.0-2.4); F2/F1 = 1.3 (1.2-1.3); APB absent on funicle; 1 PLS on each of F1-C2, 2 PLS on C3; 2-4 BPS on each of F1-C1, 1 BPS on C2 and C3; 8-9 FS on F1, 8-10 FS on F2, 8-10 FS on C1, 9 FS on C2, 7-9 FS on C3, 4-5 FS on C4; 0 US on each of F1-C3.

Genitalia: Capsule spatulate to obovate; GL/GW = 3.7 (3.4-4.1); GL/HTL = 0.6 (0.6-0.8); ADA distinctly narrower than capsule, ADA/GL = 0.4 (0.4-0.5); AI extremely pronounced, AI/GL = 0.3; PAR diverging apically from midline, with terminal spine, widest near middle, their base even with posterior edge of ADA; PAR/GL = 0.4 (0.4-0.5); VS straight and rigid, immediately adjacent to apex of genital capsule for most of its length, VS/GL = 0.5 (0.4-0.5); VP short and evenly tapering, VP/GL = 0.2 (0.1-0.2); transverse hinge present, immediately posterior of ADA; AP, DR absent.

Female (N=1)

Antenna: C/F = 2.7; F2/F1 = 1.4; 1 APB on F1 and F2, 0 APB on C3; 1 PLS on F1, 2 PLS on each of F2-C2, 4 PLS on C3; 3 BPS on F1, 2 BPS on F2, 4 BPS on C1, 1 BPS on C2 and C3; 0 FS on F1 and F2, 4 FS on C1, 8 FS on C2, 4 FS on C3; 1 UPP on C3; 7-9 US on each of F1-C1, 0 US on C2, 1 US on C3.

Ovipositor: Long, extending beyond posterior edge of metasoma, OL/HTL = 2.3.

Other Material Examined. - **AUSTRALIA: Tasmania:** 10km ENE of Numamara, 41.22°S, 147.24°E, 12.i - 6.ii.1983, I. D. Nauman and J. C. Cardale, MT (1♂); Ewart Creek, 41.58°S, 145.28°E, 26.i - 2.ii.1983, I. D. Nauman and J. C. Cardale, MT (2♂); **Queensland:** North Stradbroke Island, East Coast Rd., 10 km E of Dunwich, 27°27.12'S, 153°26.71'E, 12.xii.2002, JG, JM, AKO, SW wet meadow and woodland (1 ♂).

Comments. - One male paratype does not have a head. Otherwise, the type series is in good condition.

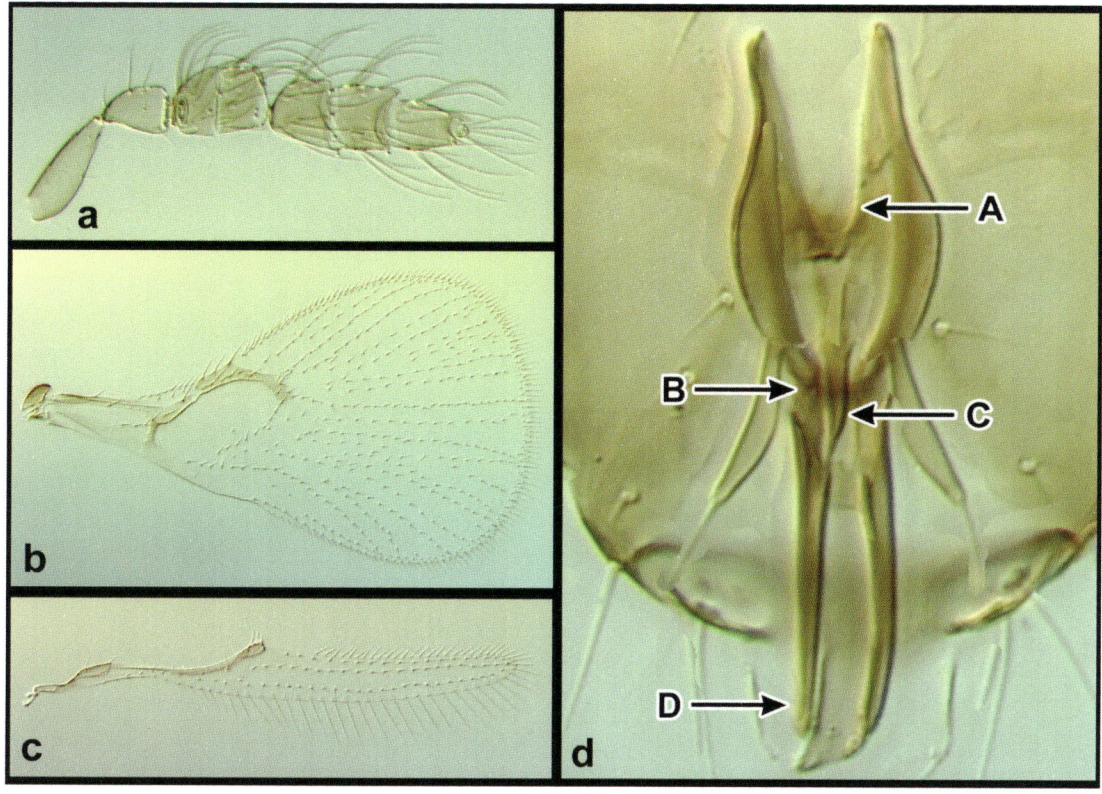

Figure 30. *Ufens invaginatus*, ♂. (a) antenna, lateral; (b) forewing, dorsal; (c) hind wing, dorsal; (d) genitalia, dorsal – arrows to {A} anterior invagination, {B} transverse hinge, {C} ventral process, {D} volsella.

Ufens kender Owen, new species
(Fig. 31)

Diagnosis. - Forewing sparsely setose with narrowly diverging setal tracks r-m to M; a single setal track between CU1 and CU2. Hind wing width not decreasing immediately apical of hamuli. Mesoscutal sculpturing longitudinally cellulate to striate. Genitalia with anterodorsal aperture somewhat 'heart' shaped and genitalic apex thickened; parameres with a terminal spine, subequal in width along entire length, their base anterior to posterior edge of anterodorsal aperture; volsellae filiform and sinuous; ventral process width at base < half of capsule width, apparently a hollow tube beginning near anterior of genital capsule.

Some specimens are known with a constriction between club segments approaching the depth found in *U. dilativena*, *U. mezentius*, *U. nazgul*, *U. pintoi*, and *U. thylacinus*. However, there is no indication that these species form a closely allied group. The genitalia of *U. kender* bear some resemblence to those of *U.*

ceratus, but can be differentiated by the presence of a ventral process, shallower anterior invagination, and shorter volsellae. There is also some resemblence between the genitalia of *U. kender* and those of species such as *U. principalis* and *U. niger. Ufens kender* is separated from these species by a deeper anterior invagination, more anterior placement of base of ventral process, and thickened area on genitalic apex.

Types. - ◆Holotype ♂ (QM). **AUSTRALIA**: **Queensland**: Mundubbera, 15 km WSW, 151°10'E, 25°38'S, 23.ix.1995, JDP, SW dry *Eucalyptus* scrub. Paratypes 3 ♂ same data (2 ♂ UCRC; 1 ♂ QM).

Etymology. - Named for Kender, a diminutive race of humanoid from the Dragonlance novels by Margeret Weis and Tracy Hickman.

Distribution. - Australia: Australian Capital Territory, New South Wales, Northern Territory, Southern Australia, Queensland, Western Australia.

Biology. - Unknown.

Description. - BL 0.7 (0.5-1.1) mm. BL/HTL = 3.6 (3.3-4.3). Mesoscutal sculpturing longitudinally cellulate to striate with interstitial sculpturing primarily transverse. Forewing sparsely setose; AA present; single setal track between CU1 and CU2; FWL/HTL = 2.9 (2.8-3.0); FWL/FWW = 1.5 (1.5-1.6); FWFS/FWW = 0.05 (0.04-0.05); Max r-m to M/Min r-m to M = 1.8 (1.5-2.3); MV/PM = 1.0 (0.9-1.1); SV/MV = 1.0 (0.9-1.2); MV length/MV width = 2.5 (2.2-3.1). Hind wing width not decreasing immediately apical of hamuli; HWL/HWW = 7.5 (6.8-8.3); HWFS/HWW = 0.7 (0.4-0.9).

Male

Antenna: Club segments somewhat loose; C/F = 2.2 (1.8-2.8); F2/F1 = 1.1 (0.9-1.4); APB absent on funicle; 1 PLS on each of F1-C2, 2 PLS on C3; 2-6 BPS on each of F1-C1, 1 BPS on C2 and C3; 8-10 FS on F1, 8-13 FS on F2, 10-12 FS on C1, 9-12 FS on C2, 8-10 FS on C3, 6-9 FS on C4; 0-1 US on F1, 0-2 US on F2, 0 US on each of C1-C3.

Genitalia: Capsule somewhat spatulate, posterior half sometimes sinuous, apex thickened; GL/GW = 2.6 (2.3-2.8); GL/HTL = 1.2 (1.0-1.3); ADA somewhat heart shaped, ADA/GL = 0.5; AI pronounced, AI/GL = 0.1 (0.07-0.1); PAR with terminal spine, subequal in width along entire length, their base distinctly anterior to posterior edge of ADA; PAR/GL = 0.4 (0.3-0.5); VP base < half capsule width, apparently a hollow tube beginning near anterior of capsule, bearing a small spine at apical quarter, VP/GL = 0.8 (0.7-0.8); transverse hinge at ca. half GL; VS filiform and sinuous, VS/GL = 0.3 (0.3-0.4); AP absent.

Female

Unknown.

Other Material Examined. - **AUSTRALIA**: **Australian Capital Territory**: Canberra, 23.iii.1981, J. R. T. Short (1♂); Blundells Creek, 3 km E Piccadilly Circus, 850 m el., 35°22S, 148°50E, iii.1985, Lawrence, Weir, Johnson, FIT (1♂); Canberra, Black Mtn., 35°16'S, 149°06'E, 30.xi-6.xii.1998, G. Gibson, YPT (1♂). **New South Wales**: nr. Moppy Lookout, Barrington Tops, 31°54'S, 151°34'E, 11.ii.1984, I. D. Naumann, ex. ethanol (1♂); Wilson R. Reserve, 15 km NW Bellangry, 7.xii.1986, D. J. Bickel, ex ethanol (1♂). **Northern Territory**: 20 km E

Humpty Doo, Fogg Dam rainforest, 31.xii.1994, S. and J. Peck, FIT (1♂); W of Alice Springs, Rd. to Ormiston Gorge, 650m. el., 23°45'02"S, 133°54'20"E, 13.iii.2002, JMH, Mallee scrub (1♂);W of Alice Springs, Ellery's Hole, 3 km E, 650 m el., 23°48'35"S, 133°11'26"E, 14.iii.2002, JMH, *Eucalyptus* creekbed (1♂); **Southern Australia**: 49 km SW Pinnaroo, 35°42'S, 140°49'E, 20 and 24.x.1983, I. D. Naumann and J. C. Cardale (1♂). **Queensland**: Braemar S. F. via Kogan, 5.ii.1980, Monteith and Raven, pyrethrum on *Geijera* (Rutaceae) (1♂); 17 km W Gamboola, 23.iv.1983, J. F. Grimshaw, SW grass near lagoon (2♂); Emerald, along Nogoa River, 13.iv.1988, JDP and G. Gordh, SW (6♂); Bauhinia, 24.8 km W, 13.iv.1988, JDP and G. Gordh, SW (1♂); Kuranda, 29.xii.1989, G. Gordh, SW rainforest (1♂); 26 km W Charleville, Rd. to Quilpie, 19.v.1991, E. C. Dahms and G. Sarnes (1♂); Cockatoo Crk. Xing, 17 km NW Heathlands, 11°39'S, 142°27'E, 22.iii-25.iv.1992, T. McLeod, MT open forest (1♂); Heathlands dump, 11°45'S, 142°35'E, 25.iv-7.vi.1992, T. McLeod, MT #2 open forest (2♂); Heathlands dump, 11°45'S, 142°35'E, 7.vi-25.vii.1992, T. McLeod, MT #2 open forest (1♂); Heathlands dump, 11°45'S, 142°35'E, 25.vii-18.viii.1992, P. Zborowski and J. Cardale, MT #2 open forest (1♂); Heathlands dump, 11°45'S, 142°35'E, 18.viii-17.ix.1992, P. Zborowski and J. Cardale, MT #2 open forest (1♂); Heathlands dump, 11°45'S, 142°35'E, 20.x-21.xi.1992, P. Zborowski and A. Calder, MT #2 open forest (1♂); Hann River, 15°11'S, 143°52'E, 24.iv-26.v.1994, P. Zborowski, MT (1♂); Auburn River NP, Auburn Falls area, 151°04'E, 25°44'S, 23.ix.1995, JDP, SW 1° *Callistemon* (2♂); Brisbane Forest Park, 27°25'04"S, 152°49'48"E, 13-19.xii.1997, MT in dry sclerophyll (2♂); Brisbane Forest Park, 27°25'04"S, 152°49'48"E, 9-16.i.1998, N. Power, MT (1♂); Brisbane Forest Park, 27°25'04"S, 152°49'48"E, 21-27.iii.1998, N. Power, MT (1♂); North Stradbroke Island, East Coast Rd., 27°26.19'S, 153°30.08'E, 12.xii.2002, JG, AKO, and JBM, SW *Banksia* scrub (2♂, 1♀); North Stradbroke Island, East Coast Rd., 10 km E Dunwich, 27°27.12'S, 153°26.71'E, 12.xii.2002, JG, AKO and JBM, SW wet meadow/woodland (1♂); Great Sandy NP, off Rainbow Beach Rd. (43), 26°00.62'S, 153°02.80'E, JBM and AKO, SW grass/*Eucalyptus* forest (1♂). **Western Australia**: Charnley River, 2 km SW Rolly Hill, Calm site 25/2, 16°22'S, 126°12'E, vi.1988, I. D. Naumann, MT open forest (1♂); South Dandalup RR crossing, 32°37'S, 116°02'E, 220m. el., C. J. Burwell, SW open forest (1♂).

Comments. - The opening for the apparently tube-like ventral process is found immediately posterior of the anterior invagination. There are several other species in which a tube-like ventral process can be inferred, e.g. the North American *U. principalis* and *U. niger*. However, it is most obvious in *U. kender*, perhaps partly due to the more anterior placement of the VP base.

This species has one of the widest known distributions among Australian species, as well as being one of the most frequently collected. The small spine in the apical portion of ventral process is most readily appreciated under SEM and is not easily seen in slide-mounted specimens. Therefore, the consistency of this trait is unknown.

Molecular data for *U. kender* were presented in Owen et al. (2007) as *Ufens* sp. 8, and can be found under Genbank accession numbers AY623537 (28S-D2+D3) and AY940378 (18S).

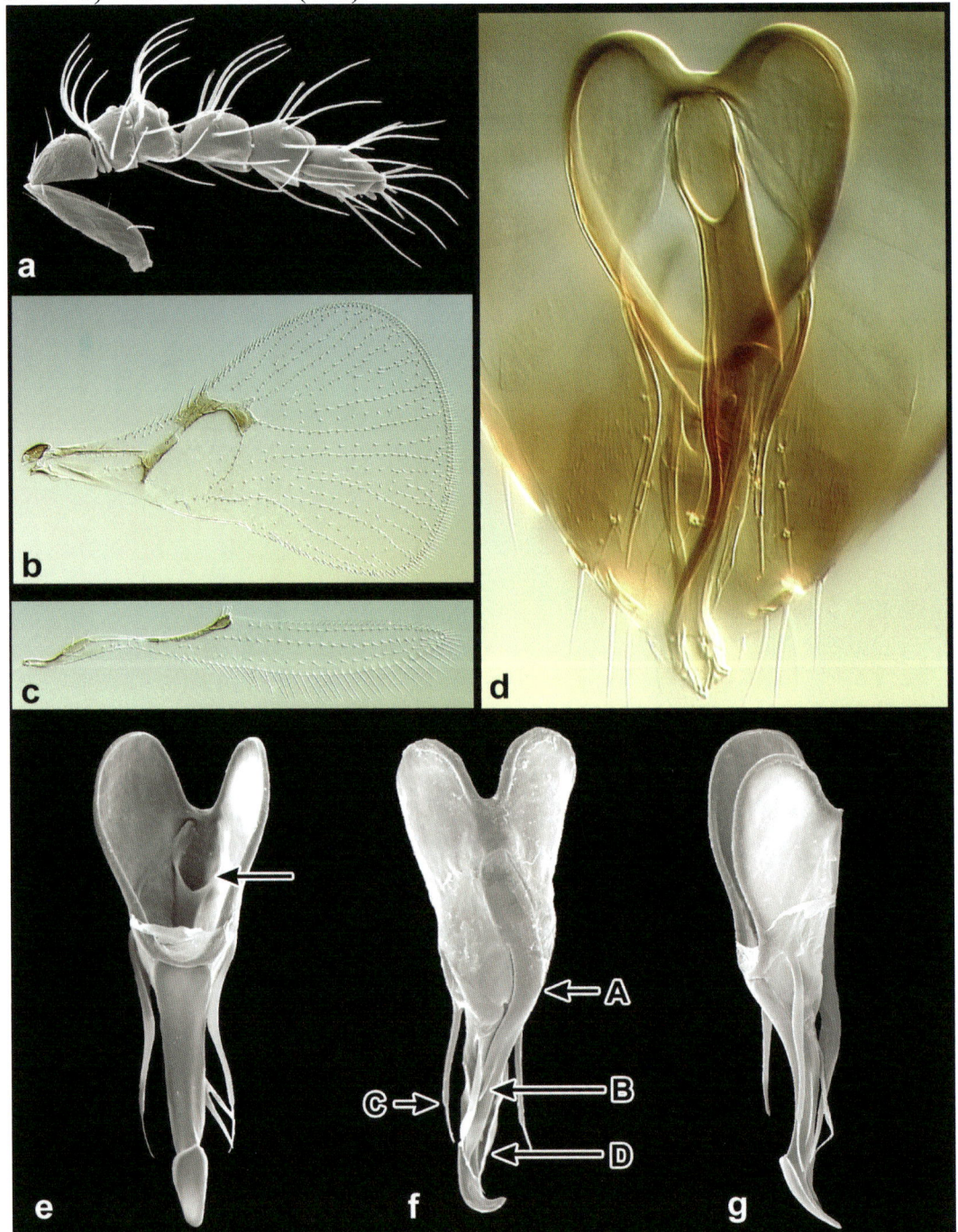

Figure 31. *Ufens kender*, ♂. (a) antenna, medial; (b) forewing, dorsal; (c) hind wing, dorsal; (d) genitalia, dorsal; (e) genitalia, dorsal – arrow to basal opening of ventral process on dorsal floor of capsule; (f) genitalia, ventral – arrows to {A} ventral process, {B} spine on ventral process, {C} paramere, {D} volsella; (g) genitalia, lateral.

Ufens khamai Owen, new species
(Fig. 32)

Diagnosis. - Forewing sparsely setose with narrowly diverging setal tracks r-m to M; a single setal track between CU1 and CU2. Hind wing width not decreasing immediately apical of hamuli. Mesoscutal sculpturing longitudinally cellulate to striate. Antenna with club segments somewhat loose. Genitalia broad anteriorly; parameres without a terminal spine, tapering, their base even with posterior edge of anterodorsal aperture, their apex with minute perforations and dramatically curving away from midline; ventral process short, its width at base > half capsule width; no other appendages present.

U. khamai is closely allied to *U. forcipis* and *U. mezentius*. It can be separated from both by the presence of a ventral process and from *U. forcipis* by the also apical curvature of the parameres. Whereas the parameres *of U. mezentius* also diverge apically from the midline, the curvature is more gradual than the abrupt deviation found in *U. khamai*. Some specimens are known with a constriction between club segments approaching the depth of *U. dilativena, U. mezentius, U. nazgul, U. pintoi,* and *U. thylacinus*. However, there is no indication that these species form a closely allied group.

Types. - ◆Holotype ♂ (CNC). **BOTSWANA: Serowe:** Farmer's Brigade, x.1987, MT, P. Forchhammer (CNC). Paratypes 13 ♂. Same data except some collected vii and ix.1987 (2♂ UCRC, 2♂ BMNH, remainder in CNC).

Etymology. - Named for King Khama, who, in 1902, moved the Ngwato people to Serowe, now Botswana's largest village and site of collection of most known specimens.

Distribution. - Botswana, India.

Biology. - Unknown.

Description. - BL 0.6 (0.5-0.7) mm. BL/HTL = 3.8 (3.4-4.2). Mesoscutal sculpturing longitudinally striate to nearly cellulate with interstitial sculpturing transverse to rugulose. Forewing sparsely setose; AA absent; single setal track between CU1 and CU2; FWL/HTL = 3.2 (3.0-3.7); FWL/FWW = 1.5 (1.4-1.5); FWFS/FWW = 0.07 (0.06-0.08); Max r-m to M/Min r-m to M = 1.8 (1.5-2.0); MV/PM = 0.9 (0.8-1.0); SV/MV = 1.2 (0.9-1.3); MV length/MV width = 1.8 (1.5-2.0). Hind wing width does not decrease immediately apical of hamuli; HWL/HWW =7.0 (6.2-8.0); HWFS/HWW = 0.8 (0.7-0.9).

Male

Antenna: Club segments somewhat loose; C/F = 2.3 (2.1-2.5); F2/F1 = 1.1 (1.0-1.1); APB absent on funicle; 1 PLS on each of F1-C2, 2 PLS on C3; 1-4 BPS on each of F1-C1, 1 BPS on C2 and C3; 6-9 FS on F1, 9-13 FS on F2, 9-11 FS on C1, 9-13 FS on C2, 7-11 FS on C3, 4-9 FS on C4; US absent on each of F1-C3.

Genitalia: Capsule broad anteriorly; GL/GW = 1.5 (1.3-1.7); GL/HTL = 0.7 (0.6-0.9); ADA broad, ADA/GL = 0.5; PAR without terminal spine, their width greatest at base, their base even with posterior edge of ADA, their apex abruptly curving away from midline at nearly 90° angle and with minute perforations; PAR/GL = 0.5

(0.4-0.5); VP difficult to discern, its base > half capsule width, VP/GL = 0.2 (0.2-0.3); transverse hinge, AI, AP, DR, and VS absent.

<u>Female</u>
Unknown.

Other Material Examined. - **INDIA**: **Karnataka**: Bangalore, 916m, 18-24.iv.1988, K. Ghorpade (1♂).

Comments. - The ventral process of *U. khamai* can be very difficult to discern in slide-mounted specimens, though it is clearly visible in the scanning electron micrographs. The ventral process of the slide-mounted holotype is easier to distinguish than in most other specimens. The disjunct distribution of this species is noteworthy, though may simply be an artifact of the relavitely few collections of this species.

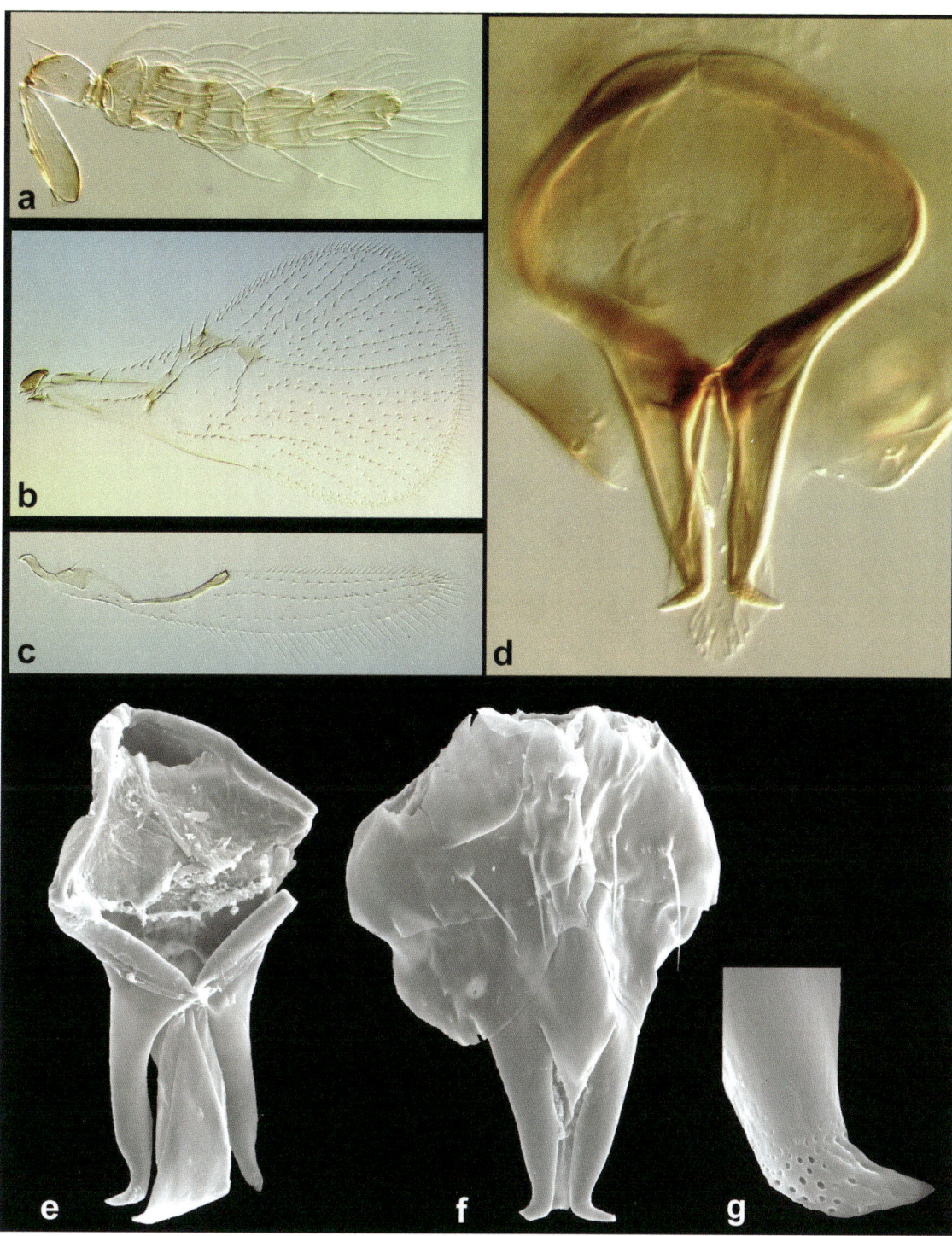

Figure 32. *Ufens khamai*, ♂. (a) antenna, lateral; (b) forewing, dorsal; (c) hind wing, dorsal; (d-e) genitalia, dorsal; (f) genitalia, ventral [basal portion covered by sternal sclerites]; (g) parameres, detail of dorsal apex showing minute perforations.

Ufens kurrajong Owen, new species
(Fig. 33)

Diagnosis. - Forewing sparsely setose with narrowly diverging setal tracks r-m to M; a single setal track between CU1 and CU2. Hind wing width decreasing immediately apical of hamuli. Mesoscutal sculpturing longitudinally cellulate. Genitalia with apodemes present; parameres with terminal spine, their width greatest near middle, their base even with posterior edge of anterodorsal aperture; ventral process width at base < half of capsule width, gradually tapering to a sharp point; dorsal projection curving posteroventrally; no other appendages present.

Among the species with apodemes, *U. kurrajong* is most likely confused with *U. vectis*, as they share a long dorsal projection that curves posteroventrally. However, *U. kurrajong* differs by its sparser forewing setation, genitalia with parameres widest near the middle, and a longer ventral process.

Types. - ◆Holotype ♂ (ANIC). **AUSTRALIA: Queensland**: Cockatoo Creek Xing, 17 km NW Heathlands, 11°39'S, 142°27'E, 26.i-29.ii.1992, P. Feehney, MT, open forest.

Etymology. Named for kurrajong, common name of the Australian endemic *Brachychiton populneus* (Schott and Endlicher) R. Br. (Sterculiaceae), a widespread tree of forests and woodland in Victoria, New South Wales and Queensland.

Distribution. Australia.

Biology. - Unknown.

Description. - BL 0.6 mm. BL/HTL = 3.6. Mesoscutal sculpturing longitudinally cellulate with interstitial sculpturing longitudinal. Forewing sparsely setose; AA present; single setal track between CU1 and CU2; FWL/HTL = 3.1; FWL/FWW = 1.7; FWFS/FWW = 0.07; Max r-m to M/Min r-m to M = 1.7; MV/PM = 1.2; SV/MV = 1.0; MV length/MV width = 2.7. Hind wing width decreasing immediately apical of hamuli; HWL/HWW = 9.1; HWFS/HWW = 1.0.

Male

Antenna: Club segments somewhat loose; C/F =1.4; F2/F1 = 0.9; APB absent on funicle; 1 PLS on each of F1-C2, 2 PLS on C3; 3-4 BPS on each of F1-C1, 1 BPS on C2 and C3; 6 FS on F1, 9 FS on F2, 9 FS on C1, 12 FS on C2, 7 FS on C3, 5 FS on C4; US absent on each of F1-C3.

Genitalia: Capsule with anterior margin nearly transverse; GL/GW = 3.1; GL/HTL = 0.7; ADA/GL = 0.4; PAR with terminal spine, their width greatest near middle, their base even with posterior edge of ADA; PAR/GL = 0.4; VP base < half of capsule width, gradually tapering to a sharp point, VP/GL = 0.4; AP/GL = 0.2; transverse hinge present; dorsal projection curving posteroventrally; AI, DR, VS absent.

Female

Unknown.

Other Material Examined. - None.

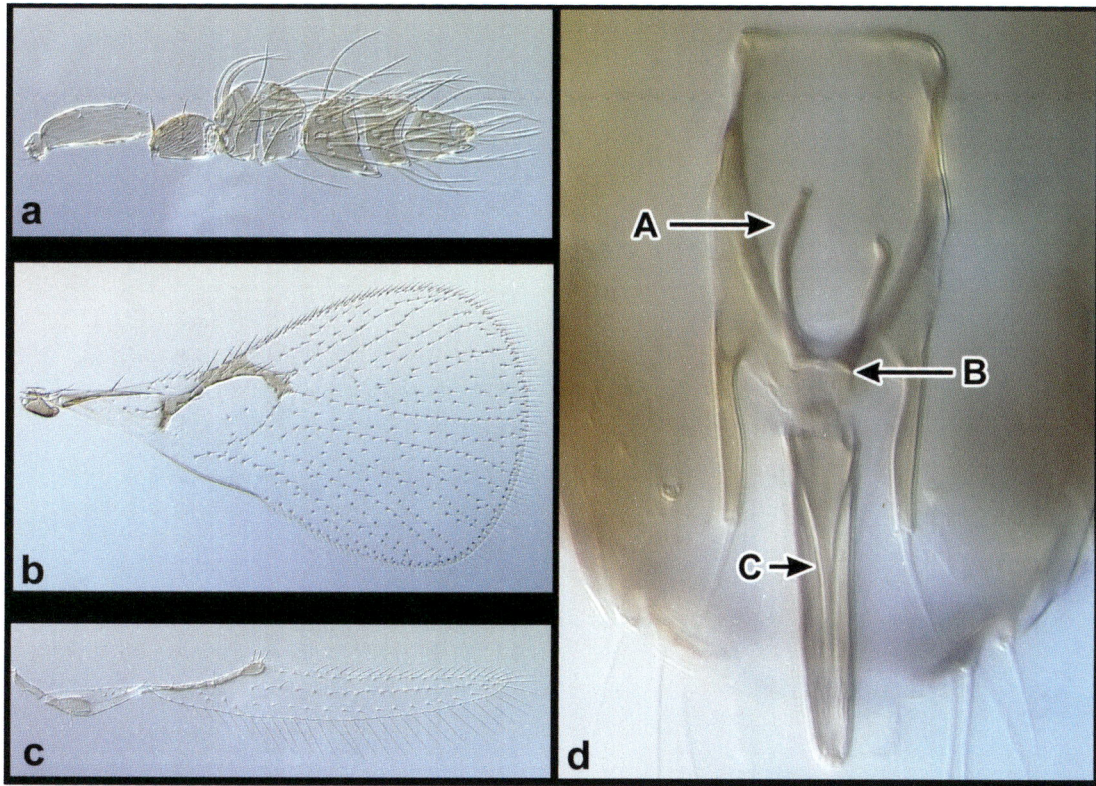

Figure 33. *Ufens kurrajong*, ♂. (a) antenna, lateral; (b) forewing, dorsal; (c) hind wing, dorsal; (d) genitalia, dorsal – arrows to {A} apodeme, {B} transverse hinge, {C} ventral process.

Ufens lanna Owen, new species
(Fig. 34)

Diagnosis. - Forewing sparsely setose with narrowly diverging setal tracks r-m to M; a single setal track between CU1 and CU2. Hind wing width not decreasing immediately apical of hamuli. Mesoscutal sculpturing longitudinally cellulate. Genitalia possessing parameres with long terminal spine, their width greatest near middle, their base even with posterior edge of anterodorsal aperture; volsellae small and near midline; no other appendages present.

U. *lanna* is the only species other than *U. pallidus* without two distinct setal tracks in the forewing costal cell. However, no other trait suggests that these species are closely related. Intraspecific variation cannot be appreciated in *U. lanna* to determine if the holotype is anomalus for this feature. Two other species, *U. invaginatus* and *U. kurrajong*, also possess parameres that are widest near the middle of their length and with a terminal spine. However, neither have the short volsellae of *U. lanna*; in *U. invaginatus* they are ca. 0.5 the genitalia length and *U. kurrajong* lacks volsellae altogether.

Types. - ♦Holotype ♂ (USNM). **THAILAND**: Chiang Mai, 20-23.iv.1989, G. T. Baker.

Etymology. - Named for the ancient Lanna Kingdom, of which Chiang Mai became capital in 1296.

Distribution. - Thailand.

Biology. - Unknown.

Description (N=1). - BL 0.8 mm. BL/HTL = 3.6. Mesoscutal sculpturing longitudinally striate approaching cellulate, with interstitial sculpturing transverse. Forewing sparsely setose, with a single setal track in costal cell; AA absent; single setal track between CU1 and CU2; FWL/HTL = 2.7; FWL/FWW = 1.5; FWFS/FWW = 0.07; Max r-m to M/Min r-m to M = 1.5; MV/PM = 0.9; SV/MV = 1.5; MV length/MV width = 2.7. Hind wing width not decreasing immediately apical of hamuli; HWL/HWW = 6.3; HWFS/HWW = 0.6.

Male

Antenna: Club segments somewhat loose; C/F = 2.4; F2/F1 = 1.5; APB absent on funicle; 1 PLS on each of F1-C2, 2 PLS on C3; 2-5 BPS on each of F1-C1, 1 BPS on C2 and C3; 10 FS on F1, 13 FS on F2, 13 FS on C1, 13 FS on C2, 9 FS on C3, 10 FS on C4; US absent on each of F1-C3.

Genitalia: Capsule narrow, with anterior margin nearly transverse; GL/GW = 7.3; GL/HTL = 0.6; ADA/GL = 0.5; PAR with long terminal spine, their width greatest near middle, their base even with posterior edge of ADA; PAR/GL = 0.4; VS near midline and short, VS/GL = 0.1; AI, AP, DR, VP, transverse hinge absent.

Female

Unknown.

Other Material Examined. - None.

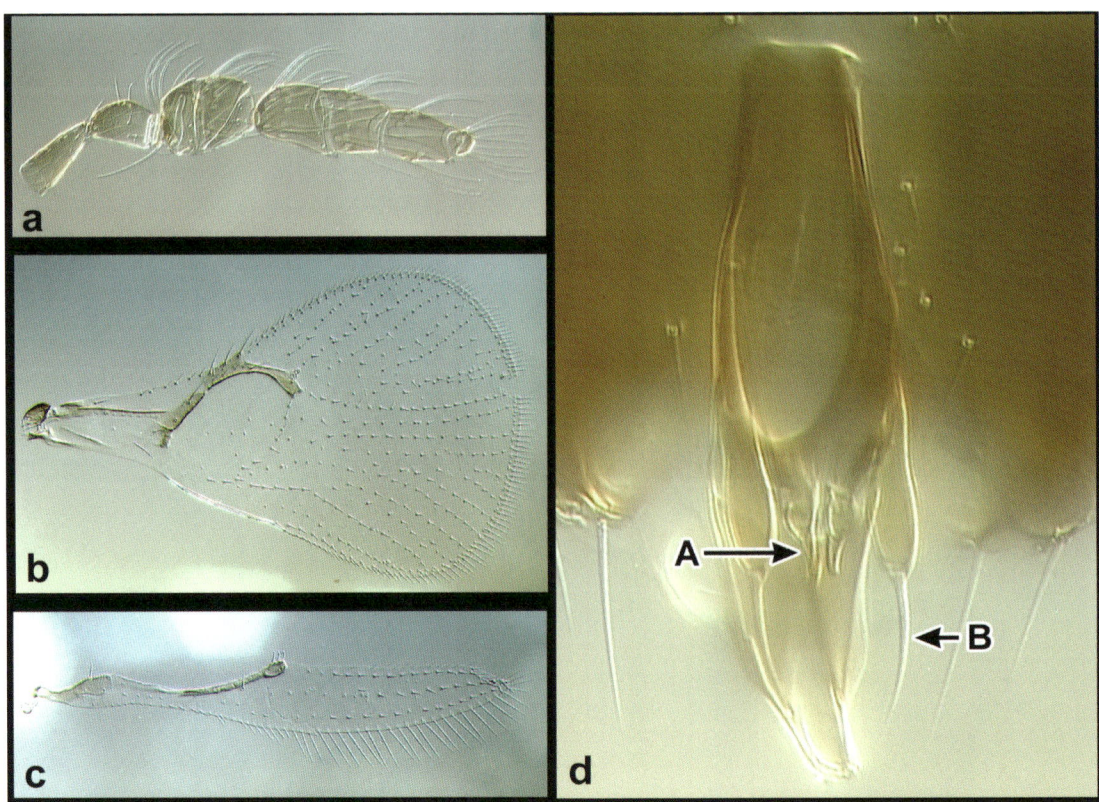

Figure 34. *Ufens lanna*, ♂. (a) antenna, lateral; (b) forewing, dorsal; (c) hind wing, dorsal; (d) genitalia, dorsal – arrows to {A} volsella, {B} long terminal spine of paramere.

Ufens messapus Owen, new species
(Fig. 35)

Diagnosis. - Forewing sparsely setose with narrowly diverging setal tracks r-m to M; a single setal track between CU1 and CU2. Hind wing width decreasing immediately apical of hamuli. Mesoscutal sculpturing longitudinally striate. Genitalia with volsellae joined at base, hooked at apex; no other appendages present.

      The elliptical capsule, broad transerve hinge, volsellae joined by a transverse bridge and hooked apically, and lack of other appendages distinguish the genitalia of *U. messapus*; they are unlikely to be mistaken for those of any other species. Perhaps the only species it may be confused with is *U. spicifer*. Both have a very compact club with a very small C4. However, *U. messapus* does not share the long spine-like volsellae and has a dorsal ridge.

Types. - ◆Holotype ♂ (QM). **AUSTRALIA: Queensland**: Emerald, along Nogoa River, 13.iv.1988, JDP and G. Gordh, SW. Paratype ♂, same data (UCRC).

Etymology. - Messapus, together with Ufens and Mezentius, were the main leaders of Latium in Virgil's Aeneid.

Distribution. - Australia.

Biology. - Unknown.

Description (N=2). - BL 0.5 (0.5) mm. BL/HTL = 3.4 (3.3-3.4). Mesoscutal sculpturing narrowly longitudinally striate with interstitial sculpturing rugulose. Forewing sparsely setose; AA present; single setal track between CU1 and CU2; FWL/HTL = 3.0; FWL/FWW = 1.8; FWFS/FWW = 0.1; Max r-m to M/Min r-m to M = 1.6 (1.5-1.6); MV long, MV/PM = 1.6 (1.6-1.7); SV/MV = 0.7; MV length/MV width = 4.3 (4.1-4.4). Hind wing width decreasing immediately apical of hamuli; HWL/HWW = 9.6 (9.3-9.8); hind wing fringe long, HWFS/HWW = 1.5 (1.4-1.5).

Male

Antenna: Club segments compact, C4 very small, not extending beyond PLS on C3; C/F = 2.3 (2.2-2.3); F2/F1 = 1.3; 1 APB on F1, 0 APB on F2; 1 PLS on each of F1-C2, 2 PLS on C3; 0 BPS on F1, 1-3 BPS on each of F2-C1, 1 BPS on C2 and C3; 0 FS on F1, 3-4 FS on F2, 5-6 FS on C1, 6-8 FS on C2, 4 FS on C3, 2 FS on C4; 0-1 US on F1, 1-3 US on F2, 2-4 US on C3.

Genitalia: Capsule long, somewhat elliptical; GL/GW = 3.1 (3.0-3.2); GL/HTL = 1.1; ADA/GL = 0.5; AI shallow, AI/GL = 0.04 (0.04-0.05); transverse hinge located at ca. half GL; VS basally joined by a transverse bridge, apparently with hooked apices, VS/GL = 0.4; DR present but obsolescent; AP, PAR, VP absent.

Female

Unknown.

Other Material Examined. - None.

Comments. - Homology of the single pair of appendages referred to as volsellae is based on their medial placement and transversely joined bases. They are reminiscient of similar structures found in *U. nazgul*, which are convincingly identified as volsellae in that species due to the presence of parameres.

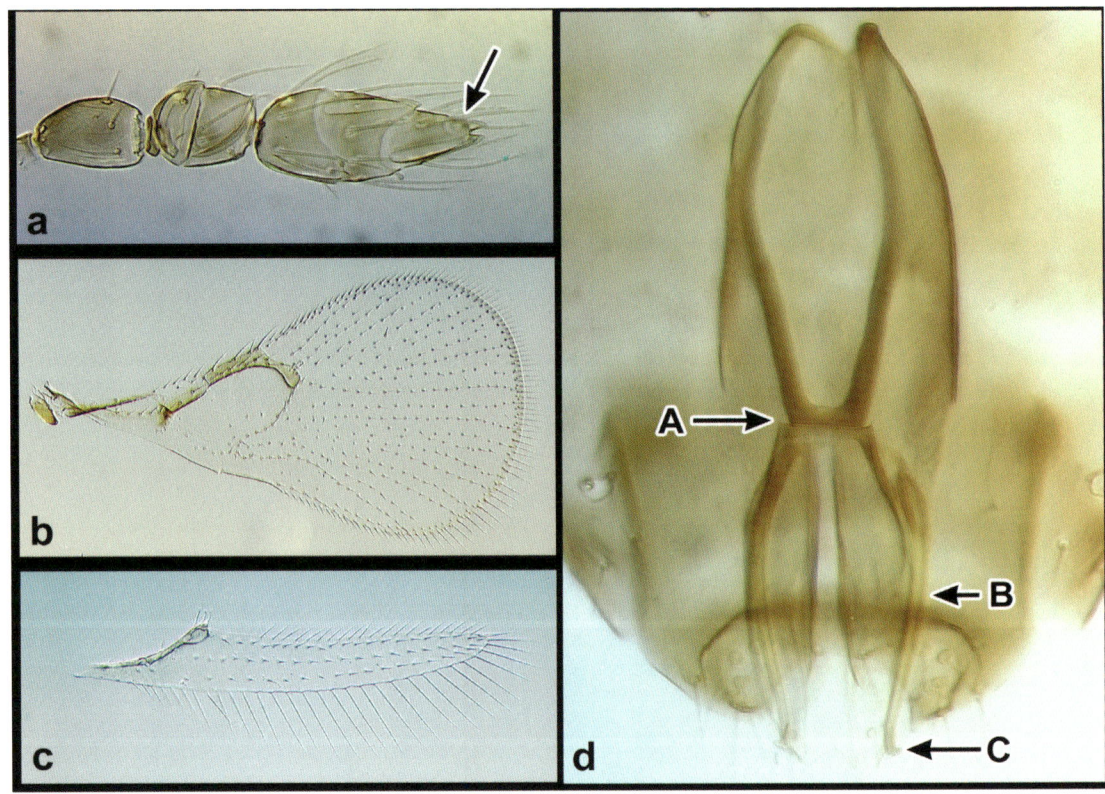

Figure 35. *Ufens messapus*, ♂. (a) antenna, lateral – arrow to small terminal club segment; (b) forewing, dorsal; (c) hind wing, dorsal; (d) genitalia, dorsal – arrows to {A} transverse hinge, {B} volsella, {C} small hook at apex of volsella.

Ufens mezentius Owen, new species
(Fig. 36)

Diagnosis. - Forewing sparsely setose with narrowly diverging setal tracks r-m to M; a single setal track between CU1 and CU2. Hind wing width not decreasing immediately apical of hamuli. Mesoscutal sculpturing longitudinally cellulate. Genitalia with capsule spatulate in outline; parameres long, without terminal spine, their width greatest at base, their base even with posterior edge of anterodorsal aperture, posteriorly gradually diverging from midline; no other appendages present.

     This species is most likely to be confused with *U. forcipis*, from which it differs by having a narrower anterodorsal aperture and parameres that more dramatically diverge apically from midline. *U. mezentius* is also potentially confused with other species whose parameres lack a terminal spine that apically diverges from the midline, including *U. cupuliformis* and *U. khamai*. *Ufens mezentius* is separated from *U. cupuliformis* by its relatively long parameres that are not laterally emarginated, as well as by its more highly convex anterior margin. *U. mezentius* is separated from *U. khamai* by its lack of a ventral process and having parameres that more gradually diverge from the midline.

Types. ◆Holotype ♂, Allotype ♀ (CNC). **ISRAEL**: **Negev**: Elat Mountain Nat. Res., Wadi Mapalim, 1,280' el, nr. Mt. Shelomo, 29°34'45"N, 34°53'38"E, M. T., 15-22.iv.1996, M. E. Irwin. Paratypes 3♂, 1♀. 1♂ same data (UCRC). Remaining with same data except as follows: 2♂ collected 22-29.iv.1996, 1♀ collected 1-6.v.1996 (UCRC).

Etymology. - Mezentius, together with Ufens and Messapus, were the main leaders of Latium in Virgil's Aeneid.

Distribution. - Israel, Sri Lanka, South Africa.

Biology. - Unknown.

Description. BL 0.7 (0.6-0.8) mm. BL/HTL = 3.8 (3.4-4.4). Mesoscutal sculpturing longitudinally striate with interstitial sculpturing transverse to rugulose. Forewing sparsely setose, AA absent, single setal track between CU1 and CU2; FWL/HTL = 3.1 (2.8-3.6); FWL/FWW = 1.5 (1.4-1.6); FWFS/FWW = 0.08 (0.05-0.2); Max r-m to M/Min r-m to M = 2.0 (1.9-2.1); MV/PM = 1.0 (0.9-1.2); SV/MV = 1.0 (0.8-1.1); MV length/MV width = 2.2 (2.0-2.4). Hind wing width does not decrease immediately apical of hamuli; HWL/HWW = 7.6 (7.5-7.9); HWFS/HWW = 0.8 (0.7-1.0).

Male

Antenna: Club segments separated by deep constriction; C/F = 2.6 (2.3-2.8); F2/F1 = 1.3 (1.2-1.4); APB absent on funicle; 1 PLS on each of F1-F2, 1-2 PLS on C1, 1 PLS on C2, 1-2 PLS on C3; 2-5 BPS on each of F1-C1, 1 BPS on C2 and C3; 6-9 FS on F1, 8-12 FS on F2, 8-13 FS on C1, 9-12 FS on C2, 8-9 FS on C3, 5-8 FS on C4; US absent on each of F1-C3.

Genitalia: Capsule somewhat spatulate in outline; GL/GW = 2.0 (1.9-2.2); GL/HTL = 0.9 (0.8-1.0); ADA/GL = 0.4 (0.3-0.4); AI very shallow, AI/GL = 0.02 (0.01-0.02); PAR without terminal spine, their width greatest at base, their base even with posterior edge of ADA, generally straight but gradually diverging from midline near apex, PAR/GL = 0.4 (0.4-0.5), PAR/PAR width = 9.0; AP, DR, VS, VP, transverse hinge absent.

Female (N=2)

Antenna: C/F = 1.9 (1.7-2.1); F2/F1 = 2.0 (1.8-2.2); 1 APB on F1 and F2, 0 APB on C3; 1 PLS on F1, 6-7 PLS on F2, 2 PLS on C1, 1-2 PLS on C2, 4 PLS on C3; 4-6 BPS on F1, 3-4 BPS on F2 and C1, 1 BPS on C2 and C3; 0 FS on F1 and F2, 7-9 FS on C1, 10-11 FS on C2, 2-3 FS on C3; 1 UPP on C3; 9-11 US on F1, 6 US on F2, 6-9 US on C1, 0 US on C2, 3-4 US on C3.

Ovipositor: OL/HTL = 1.2.

Other Material Examined. - **SRI LANKA: Monerogala District**: Buttala Udugama, ca. 25 km NE, 10.ix.1993, SW "jungle + bush bordering jungle", M.

Söderlund (1♂). **SOUTH AFRICA:** Rhenosterport Nature Reserve, 25°44.835'S, 28°58.540'E (1♂).

Comments. - Females of this species have antennae with among the most placoid sensilla on the second funicle segment of any *Ufens* species.

Some specimens of the type series are labeled Southern District, rather than Negev, but with all other information identical. Although Negev does cover the greatest amount of territory, the Southern District of Israel also is officially composed of several other towns and cities.

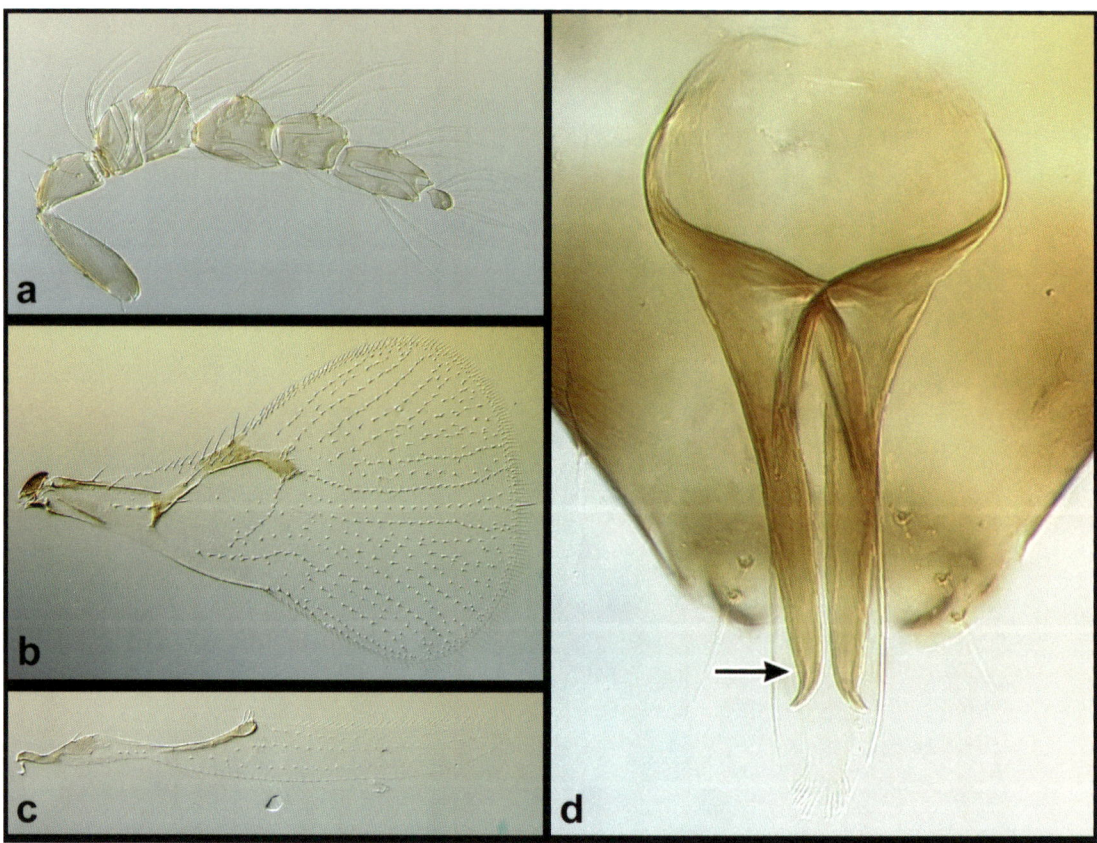

Figure 36. *Ufens mezentius*, ♂. (a) antenna, lateral; (b) forewing, dorsal; (c) hind wing, dorsal; (d) genitalia, dorsal – arrow to paramere.

Ufens mirabilis Owen, new species
(Fig. 37)

Diagnosis. - Forewing sparsely setose with narrowly diverging setal tracks r-m to M; a single setal track between CU1 and CU2. Hind wing width not decreasing immediately apical of hamuli. Mesoscutal sculpturing longitudinally striate. Genitalia with capsule broad throughout its length; parameres wide, with a thick and dark terminal spine and subequal in width along entire length, their base even with posterior edge of anterodorsal aperture; volsellae straight and thick, length subequal to that of parameres.

The terminal spine of the parameres in this species is unique: it is dark and nearly as thick as the rest of the paramere. The only other *Ufens* known to have darkened and more heavily sclerotized paramere tips is the North American *U. debachi*, a species that does not have a terminal spine.

Types. ♦Holotype ♂ (QM). **AUSTRALIA: Queensland**: Charleville, 11.7 km on Rd. to Cunnamulla, 7.iii.1989, E. Dahms and G. Sarnes, SW *Acacia murrayana* F. Muell ex Benth, *Triodia marginata* NT Burbidge, and broom brush (QM).

Etymology. - Latin for wonderful, extraordinary, unusual.

Distribution. - Australia.

Biology. - Unknown.

Description. - BL 0.9 mm; BL/HTL = 4.3. Mesoscutal sculpturing longitudinally striate with interstitial sculpturing primarily longitudinal. Forewing sparsely setose, AA present, single setal track between CU1 and CU2; FWL/HTL = 3.0; FWL/FWW = 1.7; FWFS/FWW = 0.05; Max r-m to M/Min r-m to M = 1.6; MV/PM = 0.9; SV/MV = 1.1; MV length/MV width = 2.5. Hind wing width not decreasing immediately apical of hamuli; Hind wing broad, HWL/HWW = 3.9; HWFS/HWW = 0.7.

Male (N=1)

Antenna: C/F = 2.5; F2/F1 = 1.3; APB absent on funicle; 1 PLS on each of F1-C2, 2 PLS on C3; 2-3 BPS on each of F1-C1, 1 BPS on C2 and C3; 8 FS on F1, 11 FS on F2, 11 FS on C1, 11 FS on C2, 10 FS on C3, 8 FS on C4; US absent on F1-C3.

Genitalia: Capsule with anterior margin nearly transverse, and broad throughout its length; GL/GW = 2.1; GL/HTL = 1.0; ADA/GL = 0.5; PAR wide, subequal in width along entire length, base even with posterior edge of ADA, apical portions, especially terminal spine darker (potentially more sclerotized), terminal spine nearly as wide as remainder of PAR; PAR/GL = 0.4; VP thin, base < half of capsule width, evenly tapering, VP/GL = 0.5; VS straight and thick, VS/GL = 0.4; AI extremely shallow, AI/GL = 0.01; DR extending ca. half of GL and reaching base of VP; transverse hinge at ca. 0.6 of GL; AP absent.

Female

Unknown.

Other Material Examined. - None.

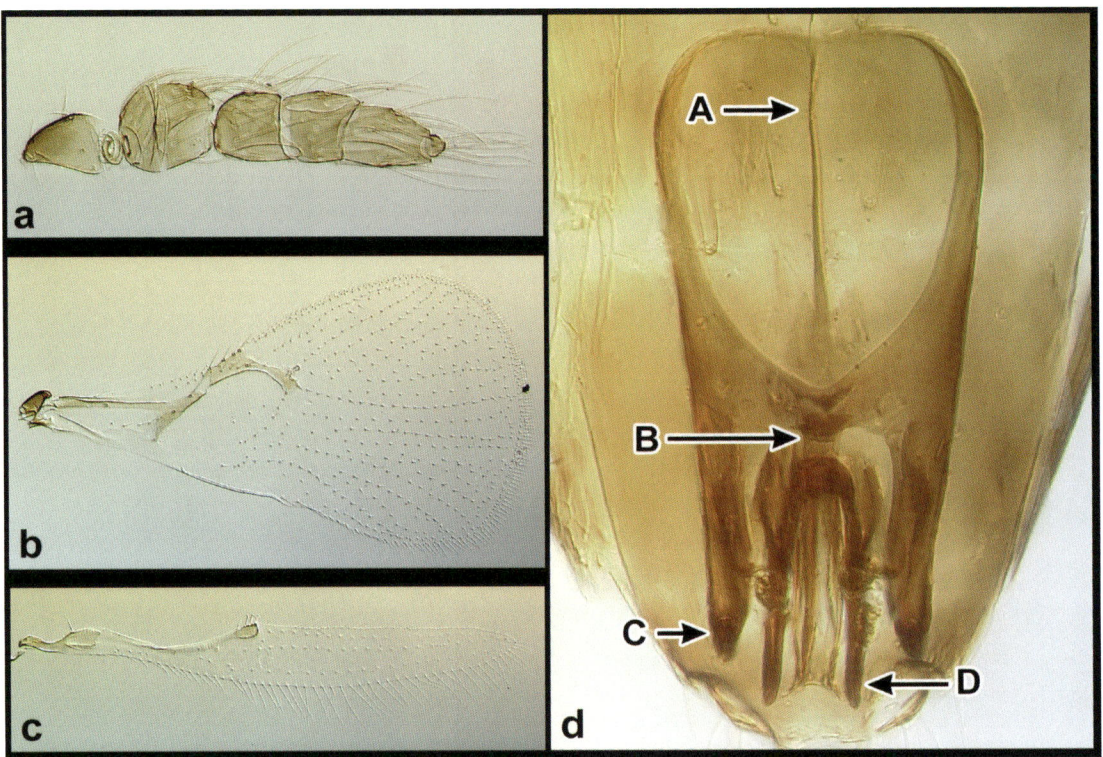

Figure 37. *Ufens mirabilis*, ♂. (a) antenna, lateral; (b) forewing, dorsal; (c) hind wing, dorsal; (d) genitalia, dorsal – arrows to {A} dorsal ridge, {B} transverse hinge, {C} thick terminal spine of paramere, {D} volsella.

Ufens nazgul Owen, new species
(Fig. 38)

Diagnosis. - Antenna club with pronounced constrictions between segments. Forewing sparsely setose with narrowly diverging setal tracks r-m to M; a single setal track between CU1 and CU2. Hind wing width increasing apical of hamuli. Mesoscutal sculpturing longitudinally striate, approaching cellulate. Genitalia with parameres short and dorsoventrally flattened, without terminal spine, widest near middle, and located rather posteriorly; volsellae difficult to discern, joined at base, and following trajectory of parameres; transverse hinge located rather apically, followed by bulbous area.

   The only other species known to have basally joined volsellae is *U. messapus*, though the species are unlikely to be confused as *U. messapus* lacks parameres and a ventral process.

Types. - ♦Holotype ♂ (QM). **AUSTRALIA: Queensland**: Rd. to Gundare, 62 km N of Charleville, 9.iii.1989, E. Dahms and G. Sarnes, Brigalow belt, SW *Acacia* sp. Paratype ♂, same data but collected 10.iii.1989 (UCRC).

Etymology. - Named for the Nazgul, or Ringwraith, from The Lord of the Rings trilogy by J. R. R. Tolkien.

Distribution. - Australia.

Biology. - Unknown.

Description (N=4). - BL 0.9 (0.8-1.0) mm; BL/HTL = 3.8 (3.6-4.0). Mesoscutal sculpturing longitudinally striate, approaching cellulate, with interstitial sculpturing primarily transverse. Forewing sparsely setose, AA absent, single setal track between CU1 and CU2; FWL/HTL = 2.9 (2.8-3.0); FWL/FWW = 1.5; FWFS/FWW = 0.03 (0.02-0.03); Max r-m to M/Min r-m to M = 1.9 (1.5-2.3); MV/PM = 0.8 (0.8-0.9); SV/MV = 1.2 (1.2-1.3); MV length/MV width = 1.9 (1.8-2.0). Hind wing width increasing beyond hamuli; HWL/HWW = 6.6 (5.6-7.3); HWFS/HWW = 0.5 (0.4-0.6).

Male

Last sternal segment distinctly sclerotized and could be mistaken for genitalic structure; with deep emargination forming an inverted 'V' ventral of capsule.

Antenna: Club segments separated by deep constriction, club comparatively long, C/F = 3.1 (2.8-3.6); F2/F1 = 1.2 (1.0-1.3); APB absent on funicle; 1 PLS on each of F1-C2, 2 PLS on C3; 4-5 BPS on each of F1-C1, 1 BPS on C2 and C3; 11-14 FS on F1, 13-19 FS on F2, 14-18 FS on C1, 13-20 FS on C2, 13-18 FS on C3, 13-15 FS on C4; US absent on F1-C3.

Genitalia: GL/GW = 2.6 (2.5-2.7); GL/HTL = 1.2 (1.2-1.3); ADA/GL = 0.5 (0.4-0.6); PAR without terminal spine, dorsoventrally flattened, widest near middle, base posterior to posterior edge of ADA; PAR/GL = 0.2 (0.2-0.3); VS arising medially and joined baasally, with trajectory following parameres, VS/GL = 0.4; VP base < half of capsule width, asymmetrically bifurcate with one side of bifurcation nearly reaching apex of capsule, VP long, VP/GL = 0.7 (0.7-0.8); AI extremely shallow, AI/GL = 0.01; transverse hinge at ca. apical 2/3 of GL, followed posteriorly by a bulbous area; AP, DR presumably absent.

Female

Unknown.

Other Material Examined. - **AUSTRALIA: New South Wales**: Fowlers Gap Research Station, 31°05'S, 141°42'E, 8-9.xii.1982, J. C. Cardale, ex. alcohol (1♂) (ANIC). **Queensland**: Taroom, 5-10 km S, 14.iv.1988, JDP and G. Gordh, SW dry *Eucalyptus* scrub (1♂) (QM).

Comments

This species has particularly pronounced constrictions between the club segments. Also, C4 has an unusually high number of flagelliform seta, and not surprisingly it appears slightly larger than C4 of other species. *U. nazgul* has bizarre genitalia that are somewhat difficult to interpret. It does not appear to have a dorsal ridge as found in many other species, though there is a broad darkened area in the same location in some specimens. This area is not considered homologous as it is indistinguishable from the base of the ventral projection. Additionally, this darkened area is lighter medially in some specimens, and is broader than the dorsal ridge of other species. The terminal spine on the parameres appears to be absent. In one specimen, however, a faint line can be observed at the paramere apex. The presence or absence of this structure requires verification by SEM examination. The volsellae are difficult to distinguish. Apically they largely overlay the parameres; they can be differentiated basally but this depends on the preparation.

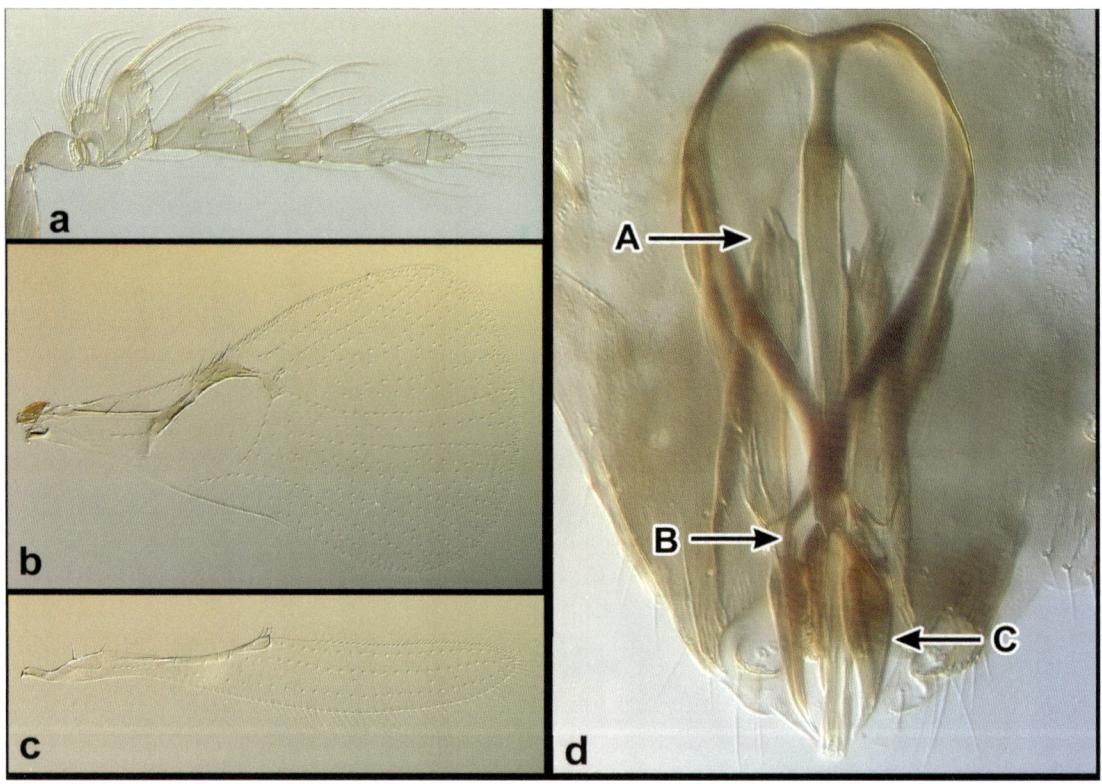

Figure 38. *Ufens nazgul*, ♂. (a) antenna, lateral; (b) forewing, dorsal; (c) hind wing, dorsal; (d) genitalia, dorsal – arrows to {A} sclerotized last sternal segment (not part of genitalia), {B} location of transverse hinge, {C} paramere.

Ufens niger (Ashmead, 1888)
(Fig. 39)

Trichogramma nigrum Ashmead, 1888: p. 107.
Ufens niger, Girault, 1911a: pp. 32-35 (generic transfer).
Doutt and Viggiani, 1968: p. 577, (forewing illustrated, Fig. 1).
Triapitsyn, 2003: pp. 251-254 (lectotype designation).
Al-Wahaibi et al. 2005: pp. 279-280 (compared with *U. ceratus* and *U. principalis*).

Diagnosis. - Forewing densely setose, with widely diverging setal tracks r-m to M; dispersed setae between CU1 and CU2; marginal vein thin. Hind wing width decreasing immediately apical of hamuli. Mesoscutal sculpturing cellulate. Genitalia with capsule anteriorly somewhat rounded and posteriorly tending to be sinuous; parameres with a terminal spine, subequal in width along entire length, their base even with posterior edge of anterodorsal aperture; volsellae anteriorly closely associated with ventral process, subequal to paramere length; ventral process evenly tapering, its base greater than half of capsule width.

Due to very similar genitalia, *U. niger* is most likely to be confused with *U. apollo, U. principalis, U. similis*, and *U. taniae*. It can be separated from all of these species by its distinctly cellulate mesoscutal sculpturing. Besides sculpturing, *U. apollo* is the most easily distinguished member of this potentially allied group as it has a sparsely setose forewing. *U. taniae* is easily separable as the only representative whose parameres lack a terminal spine. *U. niger* is separated from *U. similis* by its ventral process which tapers evenly, and is not laterally emarginate. *U. niger* can also be separated from *U. principalis* by its thinner marginal vein and more sinuous posterior half of the genitalia.

Types. - ◆Lectotype ♀ (USNM). **UNITED STATES**: "D. C.": "July 12/79", "bred from mid-rib of corn-leaves", "type No. 13396 U.S.N.M.".

Distribution. - Canada, Costa Rica, Dominican Republic, México, Puerto Rico, United States.

Biology. - According to Peck (1963), the hosts of *U. niger* include *Colladonus geminatus* (Van Duzee), *Cuerna costalis* (Fab.), *Draeculacephala mollipes* (Say), *Homalodisca* sp., *Keonella confluens* (Uhler) (Hemiptera: Cicadellidae), *Saccharosydne saccharivora* (Westwood) (Hemiptera: Delphacidae), *Diatraea* sp., *D. crambidoides* (Grote), and *D. saccharalis* (Fab.) (Lepidoptera: Crambidae). While verification of these records is difficult, it seems likely based upon what is known of other *Ufens* species that the records from Hemiptera are likely to be correct, while the records from Lepidoptera are dubious until they can be verified by further careful rearings.

Description. - BL 0.7 (0.5-0.8) mm; BL/HTL = 3.4 (2.9-3.8). Mesoscutal sculpturing cellulate, with interstitial sculpturing rugulose. Forewing densely setose; AA present; dispersed setae between CU1 and CU2; FWL/HTL = 3.0 (2.9-3.0); FWL/FWW = 1.8 (1.7-1.8); FWFS/FWW = 0.09 (0.07-0.09); Max r-m to M/Min r-m to M = 5.8 (3.7-10.0); MV/PM = 1.2 (1.1-1.3); SV/MV = 0.9 (0.8-1.0); MV narrow, MV length/MV width = 3.7 (3.2-4.4). Hind wing disk narrow, its width increasing beyond hamuli; HWL/HWW = 9.8 (9.1-11.0); HWFS/HWW = 1.2 (1.2-1.3).

Male

Antenna: Club segments somewhat compact, C/F = 2.0 (1.8-2.1); F2/F1 = 1.2 (1.0-1.3); APB absent on funicle; 1 PLS on each of F1-C2, 2 PLS on C3; 4-8 BPS on F1, 3-7 BPS on F2, 2-6 BPS on C1, 1 BPS on C2 and C3; 8-14 FS on F1, 9-15 FS on F2, 10-13 FS on C1, 10-14 FS on C2, 9-11 FS on C3, 5-11 FS on C4; US absent on F1-C3.

Genitalia: Posteriorly somewhat sinuous; GL/GW = 2.6 (2.3-2.9); GL/HTL = 1.0 (0.8-1.1); ADA/GL = 0.6 (0.5-0.7); PAR with terminal spine, subequal in width along entire length, their base roughly even with posterior edge of ADA; PAR/GL = 0.4 (0.3-0.6); VS anteriorly closely associated with VP, their length subequal to PAR length; VP width at base > half of capsule width, evenly tapering and sinuous, VP/GL = 0.7 (0.5-0.9); AI extremely shallow or absent, AI/GL = 0.01 (0-0.02); DR generally present and short, though occasionally absent; AP, transverse hinge absent.

Female (N=4)

Antenna: C/F = 2.2 (2.1-2.3); F2/F1 = 1.1 (1.0-1.5); 1 APB on F1 and F2, 1 APB on C3; 1-2 PLS on each of F1-C2, 4 PLS on C3; 3-6 BPS on F1, 4-5 BPS on F2, 2-5 BPS on C1, 1 BPS on C2 and C3; 0 FS on F1 and F2, 6-7 FS on C1, 7-11 FS on C2, 3-4 FS on C3; 1 UPP on C3; 6-8 US on F1, 6-11 US on F2, 5-7 US on C1, 0 US on C2, 3-5 US on C3.

Ovipositor: OL/HTL = 1.5 (1.2-1.8).

Other Material Examined. - **CANADA: Alberta**: Waterton Lakes NP, Prairie Field nr. picnic area on Waterton River, 14.vi.1980, I. M. Smith, YPT (1♂);Writing on Stone Prov. Park, 0.5 km E, 15-20.vii.1981, D. McCorquodale, YPT (1♂); Medicine Hat, Kin Coulee N Hwy 1, 13.vi.1982, G. Gibson (1♂); Dinosaur Prov. Park, 50°47'N, 111°30'W, 31.vii-7.viii.2000, MT in cottonwoods (2♀). **Manitoba**: Riding Mtn. NP, Dead Ox Creek, 400m el., 23.vi.1979, W. Mason, hardwood forest (4♂, 2♀). **COSTA RICA**: Puntarenas: Parque International La Amistad, Estación Altamira, sendero a Casa Coco, 1700 m el., ii.2002, C. Hanson and parataxónomos, MT (1♂, 1♀). **DOMINICAN REPUBLIC**: **Barahona**: 4 km N Paraiso, 1500m el., 22.iii.1991, L. Masner (1♂). **MÉXICO**: **Nuevo León**: Terán, 8 km N, 16.v.1984, G. Gordh (1♂). **Veracruz**: Cañon del Rio Mentlac, 3 km W Fortín de la Flores, 6.vii.1981, J. LaSalle, SW (2♂). **TRINIDAD: St. George**: Chaguaramas Bay, 16.vii.1976, J. S. Noyes, secondary forest (1♂). **UNITED STATES: Arizona**: *Cochise Co.*: Huachuca Mtns., 5364 Ash Cyn. Rd., 0.5 mi. W Hwy. 92, 5200' el., viii.1993, N. McFarland, MT (1♂). **Georgia**: *Bartow Co.*: Cartersville, 27.vi.1986, JDP, SW open field (1♂, 1♀). *Liberty Co.*: St. Catherine's Island, 30.ix-4.x.1995, A. Sharkov, SW maxi-net (3♂, 1♀) [1♂ is a molecular voucher, D#782, AY623536, AY940377]. *McIntosh Co.*: Sapelo Island, 15.ix-16.xi.1987, BRC [=CNC] Hym. Team, FIT in savanna (1♂). **Illinois**: *Champaign Co.*: Urbana, 3.ix.1983, J. T. and D. E. Huber, SW (1♂). *Effingham Co.*: 3 mi. NW Effingham, 7.vii.1980, S. Heydon (1♂). *Saline Co.*: Pankeyville, 19.vi.1991, JDP, SW riparian (1♂, 1♀). **Indiana**: *Vigo Co.*: 15.vi.1971, R. W. Meyer, ex. Hemiptera eggs (5♂, 4♀) [1 slide]. **Iowa**: *Cedar Co.*: Tipton, 12 mi. SSE, 28.viii.1983, JDP, SW (2♂, 1♀). **Louisiana**: *East Baton Rouge Parish*: Baton Rouge, private backyard, 2-4.iv.2002, S. Triapitsyn, YPT (1♂). **Maryland**: *Arundel Co.*: Patuxent Research Refuge, North tract, section S, powerline easement, 3.vii.2002, M. Gates, SW (1♂). *Montgomery Co.*: Hoyle's Mill Conservation Area, 1 mi. S White Ground, Hoyle's Mill Rds. jct., 9.vii.2003, Gates, Epstein, et al. (1♂). **Missouri**: *Wayne Co.*: Williamsville, v.1988, J. T. Becker, MT (1♂, 1♀); Williamsville, 15-30.vi.1988, J. T. Becker, MT (1♂). **South Carolina**: *Jasper Co.*: Tillman, 10 mi. NW, 26.iv.1987, L. Masner (2♂). **South Dakota**: *Charles Mix Co.*: Pickstown, 26.viii.1985, JDP, SW riparian (1♂, 1♀). **Texas**: *Brazos Co.*: College Station, Lick Creek Park, 30.vii.1987, JBW, SW (4♂); College Station, Lick Creek Park, 4-26.viii.1987, JMH and JBW, MT (1♂); College Station, Lick Creek Park, 7-10.v.1998, J. S. Noyes (1♂, 1♀). *Hidalgo Co.*: Rio Rico Rd., 2 mi. SE of Relampago, 11.vii.1985, C. W. Melton, ex. leafhopper eggs in sugar cane (2♂, 4♀). *Jeff Davis Co.*: Jeff Davis St. Park, 6.vii.1983, A. J. Mayor

(1♂, 1♀). *Jim Wells Co.*: La Copita Res. Station: North Fence Pasture 52, 23.iii.1990, G. Zolnerowich (1♂). *Montgomery Co.*: Jones State Forest, 8 mi. S Conroe, 1.iv.1987, Wharton and Carroll (1♂). *Presidio Co.*: Big Bend Ranch, 2.6 mi. E La Sauceda, 29°28'17"N, 103°54'51" W, 15.v.1990, G. Zolnerowich (2♂); Big Bend Ranch SNA, 2.6 mi. E La Sauceda, 29°28'17"N, 103°54'51" W, 18.vi.1990, JBW (1♂). *Robertson Co.*: 8 mi. E. Hearne, 3-27.x.1990, M. Hallmark, MT (1♂); 8 mi. E Hearne, 1-21.iv.1991, M. Hallmark, MT (1♂). **Tennessee**: *Knox Co.*: Knoxville, Jct. I-40 and 11E, 24.vi.1986, JDP, SW (1♂, 1♀). **Wyoming**: *Sheridan Co.*: Story, 28.vii.1983, G. Gordh, SW (1♂, 1♀).

Comments. - Ashmead (1888) claimed to have described *Trichogramma nigrum* from two specimens. These two females were remounted into Canada balsam by A. A. Girault (Girault 1911a). However, Triapitsyn (2003) was only able to locate one of these, which he designated as a lectotype. Also written on the lectotype slide "like flavipes, Paratypes [crossed out], club, 1 ring joint, n. genus [crossed out along with another word], Ufens hyalipennis [also a few other illegible words]". This lectotype is in poor condition, broken into many pieces, and the head and body are in excess balsam outside of the coverslip. Nevertheless, the forewing venation is consistent with other specimens of *U. niger*, and no other species are currently known from the northeastern United States where this specimen was collected. Therefore, in spite of the lectotype's poor condition, there is little doubt of its conspecificity with other specimens identified as *U. niger*.

Volsellae of this species can be somewhat difficult to distinguish and measure as they are very closely appressed to the remainder of the genitalia.

Molecular data for *U. niger*, as presented in Owen et al. (2007), can be found under Genbank accession numbers AY623536 (28S-D2+D3) and AY940377 (18S).

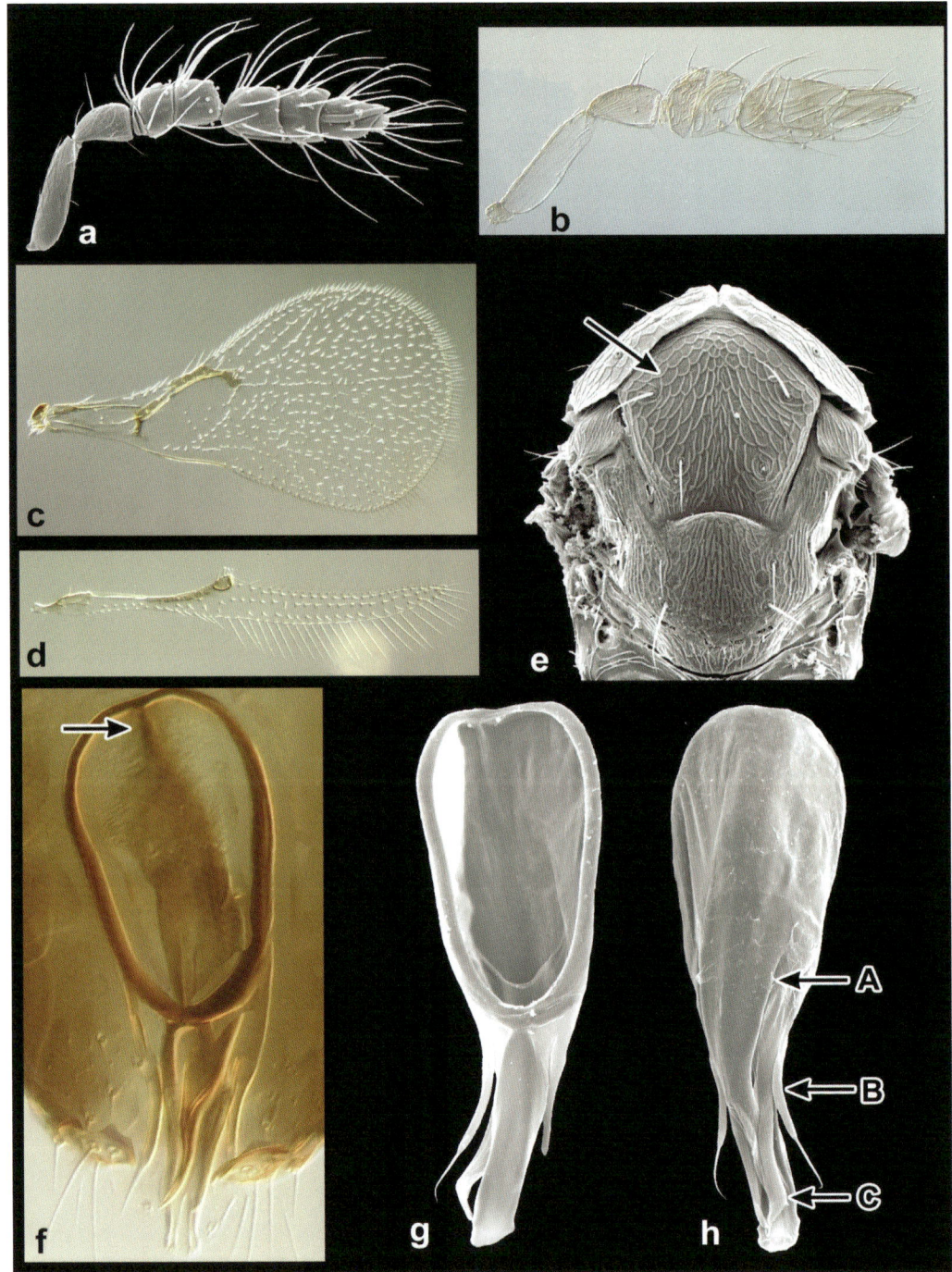

Figure 39. *Ufens niger*. (a) ♂ antenna, medial; (b) ♀ antenna, lateral; (c) forewing, dorsal; (d) hind wing, dorsal; (e) mesosoma, dorsal – arrow indicating cellulate sculpturing; (f) ♂ genitalia, dorsal – arrow to dorsal ridge; (g) ♂ genitalia, dorsal; (h) ♂ genitalia, ventral – arrows to {A} base of vental process, {B} paramere, {C} apex of volsella.

Ufens noyesi **Owen, new species**
(Fig. 40)

Diagnosis. - Forewing densely setose, with narrowly diverging setal tracks r-m to M; single setal track between CU1 and CU2. Mesoscutal sculpturing longitudinally striate. Genitalia wide and compact; anterodorsal aperture somewhat spatulate in outline; parameres with a terminal spine, subequal in width along entire length, their base even with posterior edge of anterodorsal aperture; ventral process evenly tapering, its base < half of capsule width and arising in anterior half of capsule.

This is one of the few species known with APB on the funicle. The others include *U. decipiens*, *U. gloriosus*, *U. messapus*, and *U. spicifer*. Of these, *U. messapus* and *U. spicifer* have 'normal' numbers of placoid sensilla on the funicle (1 on each segment), whereas the remainder of species have a greater number. The genitalia of *U. noyesi* do not necessarily suggest a relationship with the above species, however. The absence of volsellae and the spatulate outline of the anterodorsal aperture is shared only with *U. messapus*. In strictly genitalic terms, no other species is likely to be confused with *U. noyesi* due to its broad capsule, long and distinctively shaped anterodorsal aperture, and base of its ventral process in the posterior half of capsule.

Types. - ◆Holotype ♂ (QM). **AUSTRALIA: Western Australia**: 13 km S Norseman on Hwy 1, 29.xii.1986, J. S. Noyes.

Etymology. - Named for John S. Noyes, collector of the only known specimen of this species.

Distribution. - Australia.

Biology. - Unknown.

Description (N=1). - BL 0.6 mm; BL/HTL = 3.5. Mesoscutal sculpturing longitudinally striate, generally without interstitial sculpturing. Forewing sparsely setose; AA present; single setal track between CU1 and CU2; FWL/HTL = 2.6; FWL/FWW = 1.5; Max r-m to M/Min r-m to M = 1.5; MV/PM = 1.1; SV/MV = 1.0; MV length/MV width = 3.0. Hind wing unknown.

Male

Antenna: Club segments compact, C/F = 1.6; F2/F1 = 1.0; 1 APB F1 and F2; 6 PLS on F1, 3 PLS on F2, 1 PLS on C1 and C2, 3 PLS on C3; 3 BPS on F1, 4 BPS on F2, 2 BPS on C1, 1 BPS on C2 and C3; 9 FS on F1, 12 FS on F2, 8 FS on C1, 14 FS on C2, 11 FS on C3, 8 FS on C4; US absent on F1-C3.

Genitalia: Capsule wide and compact, GL/GW = 1.9; GL/HTL = 0.8; ADA abruptly constricted in posterior half, ADA/GL = 0.7; PAR with terminal spine, subequal in width along entire length, their base approximately even with posterior edge of ADA; PAR/GL = 0.3; AI/GL = 0.08; VP tapering, its base < half width of capsule and located in anterior half of capsule, VP/GL = 0.5; transverse hinge present; AP, DR, VS, absent.

Female
Unknown.

Other Material Examined. - None.

Comments. - The only known specimen of this species is lacking hind wings. Also, its forewings are somewhat damaged, precluding measurement of the fringe setae.

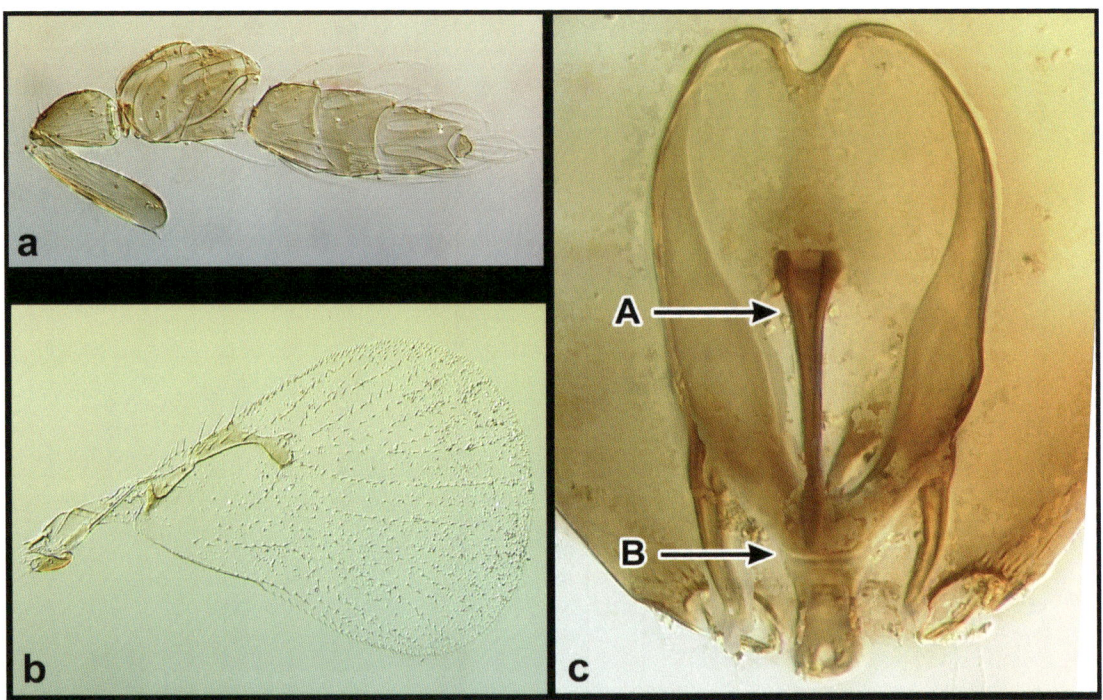

Figure 40. *Ufens noyesi*, ♂. (a) antenna, lateral; (b) forewing, dorsal; (c) genitalia, dorsal – arrows to {A} ventral process, {B} transverse hinge [Note that terminal spine of parameres are indistinct in image, but are present].

Ufens pallidus **Owen, new species**
(Fig. 41)

Diagnosis. - Forewing sparsely setose, with narrowly diverging setal tracks r-m to M; single setal track between CU1 and CU2; venation pale and base of stigmal vein broad. Hind wing width increasing apical of hamuli. Mesoscutal sculpturing longitudinally cellulate. Genitalia with anterodorsal aperture narrow; volsellae stout, crescent-shaped; no other appendages present. Females antenna with abundant placoid sensilla, especially on the third club segment; lacking unsocketed setae on the third club segment; very few basiconic peg sensilla on the funicle and club.

The forewings of this species make it easier to recognize than most. Its pale forewing venation and broad base of the stigmal vein are both unique among *Ufens* species. The genitalia are also unique. Although several species, such as *U. elimaeae*, *U. invaginatus*, and *U. rimatus* share its relatively narrow anterodorsal aperture, the crescent-shaped volsellae of *U. pallidus* are distinctive.
Types. - Holotype ♂ (USNM). **TURKMENISTAN**: Mary Region, Kara Kum desert, 15.vi.1992, S. Triapitzin, ex. *Salsola richteri*. Paratypes 5♂, 3♀; 1♂, 1♀

card mounted. 5♂, 2♀ with same data, 1♀ as above, but collected 4.vi.1992 and also collected on *Atriplex* (1♂, 1♀ BMNH, remainder (including card-mounted) UCRC).

Etymology. - Latin for pale, lacking intensity of color, in reference to the pale wing venation.

Distribution. - Turkmenistan.

Biology. - This species has been reared from unknown hosts on plants in the Chenopodiaceae, *Salsola richteri* (Moq.) Karel. ex Litv. and *Atriplex* sp.

Description. - BL 0.6 (0.5-0.6) mm; BL/HTL = 3.4 (3.2-3.6). Mesoscutal sculpturing longitudinally cellulate with interstitial sculpturing longitudinally striate to rugulose. Forewing sparsely setose, venation only lightly sclerotized; AA present; single setal track between CU1 and CU2; FWL/HTL = 3.2; FWL/FWW = 1.5 (1.5-1.6); FWFS/FWW = 0.03 (0.02-0.05); Max r-m to M/Min r-m to M = 2.4 (2.3-2.5); MV/PM = 1.1 (1.0-1.2); SV base broad, no obvious constriction between MV and SV; SV/MV = 1.0 (0.9-1.0); MV length/MV width = 1.6 (1.4-1.9). Hind wing width increasing beyond hamuli; HWL/HWW = 8.6 (7.8-9.5); HWFS/HWW = 1.0 (0.9-1.1).

Male

Antenna: Club segments somewhat compact, C/F = 1.9 (1.7-2.1); F2/F1 = 1.0 (0.7-1.2); APB absent on funicle; 2 PLS on each of F1 and F2, 1 PLS on C1, 2-3 PLS on each of C2-C3; 0-1 BPS on F1, 1 BPS on each of F2 and C1, 0-1 BPS on C2 and C3; 6-7 FS on F1, 7-9 FS on F2, 6-9 FS on C1, 8-9 FS on C2, 6-9 FS on C3, 5-6 FS on C4; US absent on F1-C3.

Genitalia: GL/GW = 3.1 (2.8-3.3); GL/HTL = 0.8 (0.8-0.9); Width of ADA distinctly narrower than capsule width, ADA/GL = 0.5 (0.4-0.5); AI somewhat indistinct, AI/GL = 0.1 (0.07-0.2); VS stout, crescent shaped, apically convergent on midline; transverse hinge present; AP, DR, PAR, VP absent.

Female (N=2)

Antenna: C/F = 1.9; F2/F1 = 1.2 (0.9-1.6); 1 APB on F1 and F2, 0 APB on C3; 2 PLS on F1, 5 PLS on F2, 2 PLS on C1, 4 PLS on C2, 6 PLS on C3; 1 BPS on each of F1-C3; 0 FS on F1 and F2, 4 FS on C1, 7 FS on C2, 4 FS on C3; 0 UPP on C3; 4-5 US on each of F1-C1, US absent on C2 and C3.

Ovipositor: OL/HTL = 1.3.

Other Material Examined. - None.

Comments. - The prominent paired appendages are inferred to be volsellae due to their medial placement. The curved nature of these structures precludes their accurate measurement. This species may have parameres, as there is some suggestion of one laterally and slightly anterior of the base of a volsella in one specimen. However, it is indistinct and there is no indication of similar structures in other specimens. It is tentatively assumed that *U. pallidus* does not have parameres, though SEM verification is needed.

Females of this species possess rather distinctive sensillar patterns, with more placoid sensilla than most, especially on the third club segment. Additionally, all species other than *U. austini* and *U. pallidus* are known to have unsocketed setae on the third club segment, and all other species have a greater number of basiconic peg sensilla on the funicle and club.

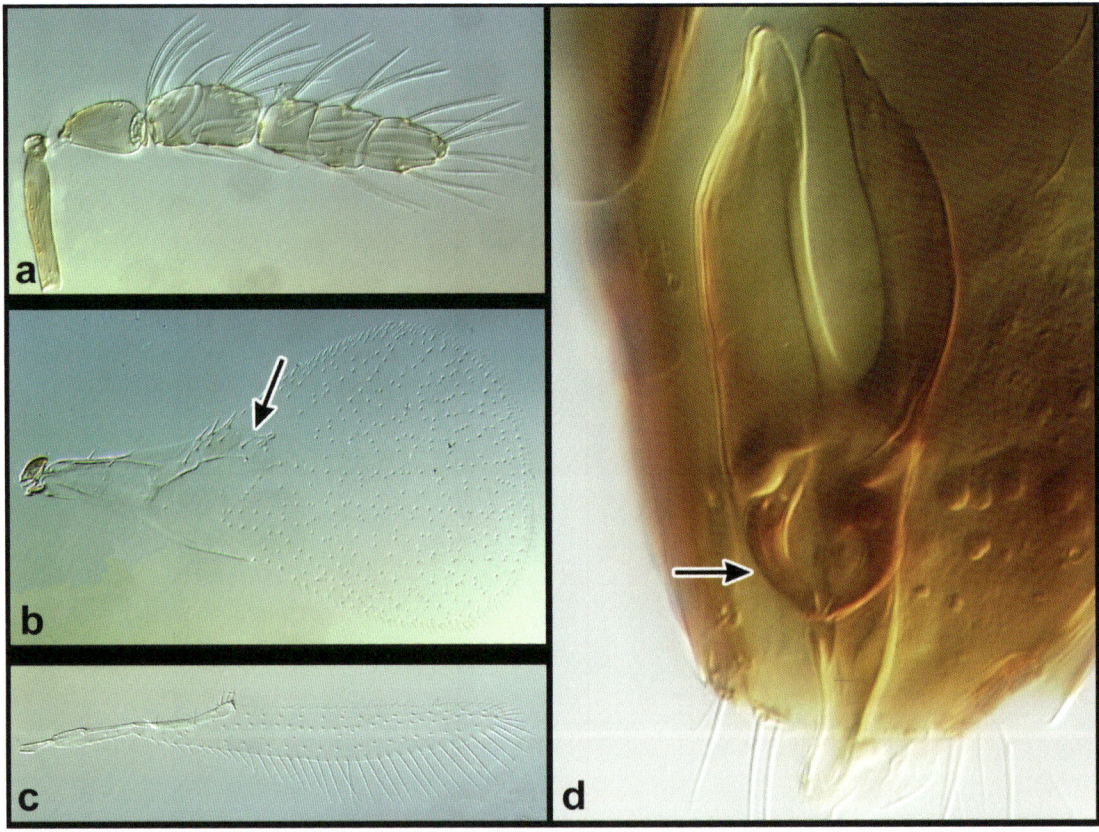

Figure 41. *Ufens pallidus*, ♂. (a) antenna, lateral; (b) forewing, dorsal – arrow to stigmal vein with broad base; (c) hind wing, dorsal; (d) genitalia, dorsal – arrow to crescent-shaped paramere.

Ufens parvimalis Owen, new species
(Fig. 42)

Diagnosis. - Antenna of male with abundant placoid sensilla on the funicle and first three segments of the club. Forewing sparsely setose, with narrowly diverging setal tracks r-m to M; single setal track between CU1 and CU2. Hind wing width increasing apical of hamuli. Mesoscutal sculpturing longitudinally cellulate to striate. Genitalia with anterodorsal aperture long; parameres with terminal spine, subequal in width along entire length, their base even with posterior edge of anterodorsal aperture; volsellae stout and rigid, their base immediately lateral of base of ventral process; ventral process width at base < half of capsule width.

    U. parvimalis has abundant placoid sensilla on all segments of the funicle and the first three of the club, a trait shared with only *U. gloriosus* and *U. placoides*. The genitalia of *U. parvimalis* are distinctive, and its lack of aedeagal apodemes separate it from the aforementioned species. *U. parvimalis* does have superficially similar genitalia to those of *U. bestiolis*. It can be separated from that species by its greater number of placoid sensilla on the antenna, lack of a dorsal ridge, and the comparatively longer and wider anterodorsal aperture.

Types. - ◆Holotype ♂ (ANIC). **AUSTRALIA: Southern Australia**: Brookfield Cons. Park, 34°21'S, 139°29'E, 24-26.xi.1992, I. D. Naumann and J. C. Cardale. Paratypes 4♂, same data (2♂ ANIC, 2♂ UCRC).

Etymology. - Latin, conjunction of parvus, meaning small and animalis, meaning animal, living, animate.

Distribution. - Australia.

Biology. - Unknown.

Description. - BL 0.8 (0.8-0.9) mm; BL/HTL = 3.6 (3.5-3.8). Mesoscutal sculpturing longitudinally cellulate approaching striate, with interstitial sculpturing transverse. Forewing sparsely setose; AA present; single setal track between CU1 and CU2; FWL/HTL = 3.0 (2.8-3.2); FWL/FWW = 1.6 (1.5-1.6); FWFS/FWW = 0.04 (0.03-0.05); Max r-m to M/Min r-m to M = 1.6 (1.3-1.8); MV/PM = 1.0 (0.9-1.1); SV/MV = 1.1 (0.9-1.2); MV length/MV width = 1.8 (1.6-2.2). Hind wing width increasing beyond hamuli; HWL/HWW = 7.1 (6.9-7.5); HWFS/HWW = 0.6 (0.6-0.7).

Male

Antenna: C/F = 1.9 (1.8-1.9); F2/F1 = 1.0 (0.8-1.2); APB absent on funicle; 4-6 PLS on F1, 2-4 PLS on F2, 2-3 PLS on C1, 3-4 PLS on each of C2-C3; 3-5 BPS on each of F1-C1, 1 BPS on C2 and C3; 4-10 FS on F1, 17-19 FS on F2, 13-18 FS on C1, 14-18 FS on C2, 5-13 FS on C3, 9-12 FS on C4; US absent on F1-C3.

Genitalia: GL/GW = 2.6 (2.3-2.7); GL/HTL = 0.6 (0.6-0.7); ADA long, ADA/GL = 0.6 (0.5-0.6); PAR with terminal spine, subequal in width along entire length, their base even with posterior edge of ADA; PAR/GL = 0.4; AI pronounced, AI/GL = 0.2 (0.1-0.2); VS straight and rigid, arising immediately lateral of base of VP, VS/GL = 0.3; transverse hinge present; VP narrow, base < half of capsule width, VP/GL = 0.3; AP, DR absent.

Female

Unknown.

Other Material Examined. - **AUSTRALIA: Northern Territory**: Georges Creek, 3 km S Ormiston Gorge, 650 m el., 23°39'S, 132° 44'E, 13.iii.2002, C. J. Burwell, SW (1♂).

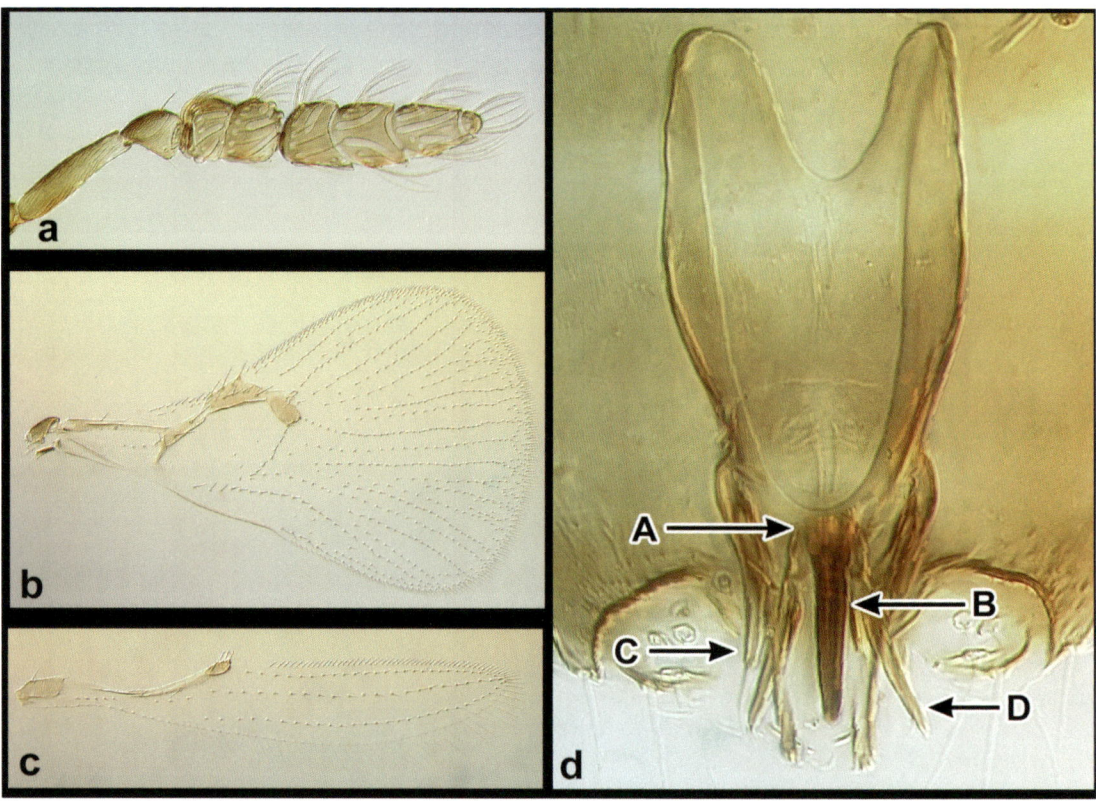

Figure 42. *Ufens parvimalis*, ♂. (a) antenna, lateral; (b) forewing, dorsal; (c) hind wing, dorsal; (d) genitalia, dorsal – arrows to {A} transverse hinge, {B} ventral process, {C} paramere, {D} volsella.

Ufens pintoi Owen, new species
(Fig. 43)

Diagnosis. - Forewing densely setose, with narrowly diverging setal tracks r-m to M; single setal track between CU1 and CU2. Hind wing width increasing apical of hamuli. Mesoscutal sculpturing longitudinally cellulate. Genitalia with apodemes present; anterodorsal aperture spatulate in outline; parameres with terminal spine, subequal in width along entire length, their base anterior to posterior edge of anterodorsal aperture; volsellae straight; ventral process width at base < half of capsule width.

      This species is most likely to be confused with *U. placoides*, as they share similar genitalic shape and aedeagal apodemes. However, *U. pintoi* does not share the abundance of placoid sensilla on the antenna, has a broader genitalic capsule, and has a more spatulate anterodorsal aperture. Although other *Ufens* species, such as *U. noyesi* and *U. thylacinus*, share a similar shape of the anterodorsal aperture, all lack aedeagal apodemes.

Types. - ◆Holotype ♂ (QM). **AUSTRALIA: Western Australia**: Stirling Range NP, ~2 km SW Camel Lake, 34°18.4'S, 118°01.1'E, 600m el., 13.xi.2002, JDP, SW.

Etymology. - Named for John D. Pinto, collector of the only known specimen.
Distribution. - Australia.
Biology. - Unknown.
Description (N=1). - BL 0.8 mm; BL/HTL = 3.9. Mesoscutal sculpturing longitudinally cellulate, with interstitial sculpturing transverse. Forewing fairly densely setose; AA present; single setal track between CU1 and CU2; FWL/HTL =2.8; FWL/FWW = 1.6; FWFS/FWW = 0.04; Max r-m to M/Min r-m to M = 2.2; MV/PM = 0.8; SV/MV = 1.2; MV length/MV width = 2.2. Hind wing width increasing beyond hamuli; Hind wing broad, HWL/HWW = 3.3; HWFS/HWW = 0.6.

Male

Antenna: Club segments loosely joined, C/F = 2.2; F2/F1 = 1.1; APB absent on funicle; 1 PLS on each of F1-C1, 2 PLS on C2 and C3; 6 BPS on each of F1-C1, 1 BPS on C2 and C3; 9 FS on F1, 12 FS on F2, 8 FS on C1, 14 FS on C2, 11 FS on C3, 8 FS on C4; US absent on F1-C3.

Genitalia: GL/GW = 2.6; GL/HTL = 1.3; ADA abruptly constricted in posterior half, ADA/GL = 0.5; PAR with terminal spine, subequal in width along entire length, their base anterior to posterior edge of ADA; PAR/GL = 0.4; AI shallow, AI/GL = 0.06; VS straight, closely appressed to capsule and primarily visible apically, VS/GL = 0.4; AP/GL = 0.3; transverse hinge present; VP thin, base <half of capsule width, VP/GL = 0.7; DR absent.

Female

Unknown.

Other Material Examined. - None.

Comments. - The volsellae of this species are somewhat difficult to discern, especially basally as this area is partially obscured by the ventral process. Nevertheless, their apices diverge slightly from the capsule.

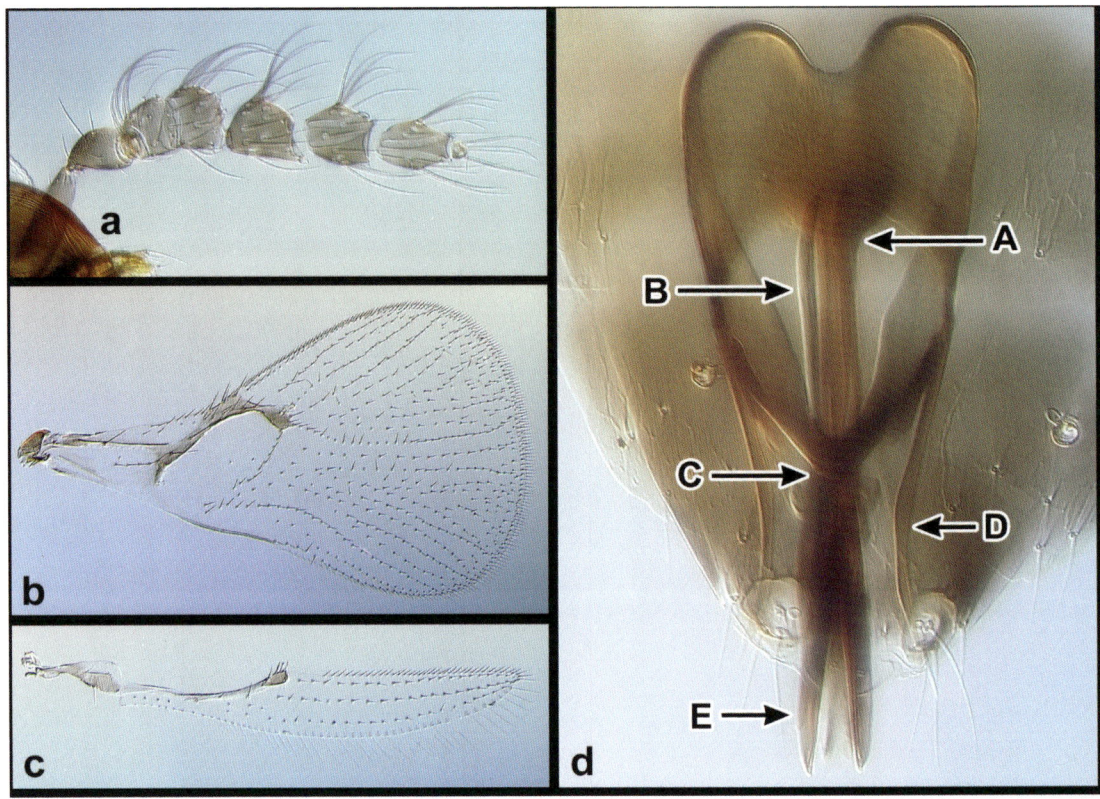

Figure 43. *Ufens pintoi*, ♂. (a) antenna, lateral; (b) forewing, dorsal; (c) hind wing, dorsal; (d) genitalia, dorsal – arrows to {A} ventral process base, {B} apodeme, {C} transverse hinge, {D} paramere, {E} volsella.

Ufens placoides Owen, new species
(Fig. 44)

Diagnosis. - Antenna of male with numerous placoid sensilla on all funicle and the first three club segments; numerous basiconic peg sensilla on the funicle and first club segment. Forewing densely setose, with narrowly diverging setal tracks r-m to M; single setal track between CU1 and CU2. Hind wing width does not decrease immediately apical of hamuli. Mesoscutal sculpturing cellulate. Genitalia with apodemes present; parameres long and with terminal spine, subequal in width along entire length, their base anterior to posterior edge of anterodorsal aperture; ventral process asymmetrically bifurcating in posterior third, with only one side of bifurcation approaching apex of capsule, width of process at base < half of capsule width; volsellae absent.

    Ufens placoides has numerous placoid sensilla on all segments of the funicle and the first three of the club, a trait shared with only *U. gloriosus* and *U. parvimalis*. The presence of parameres easily separates it from *U. parvimalis*, and the relatively straight and elongate parameres distinguish it from *U. gloriosus*. *Ufens placoides* shares a similar genitalic shape and presence of apodemes with *U. pintoi*.

However, the latter has considerably fewer placoid sensilla on the antenna, has a broader genital capsule, and has a more spatulate outline of anterodorsal aperture. *U. placoides* is also noteworthy for having more basiconic peg sensilla on the funicle and first club segments than most other species.

Types. - ◆Holotype ♂ (QM). **AUSTRALIA: Australian Capital Territory**: Canberra, Black Mtn., 35°16'S, 149°06'E, 28.xii.1998-3.i.1999, G. Gibson, MT.

Etymology. - Named in reference to the numerous placoid sensilla on the antenna of this species.

Distribution. - Australia.

Biology. - Unknown.

Description (N=3). BL 1.0 mm; BL/HTL = 4.2 (4.0-4.4). Mesoscutal sculpturing cellulate, with interstitial sculpturing rugulose to longitudinal. Forewing densely setose; AA absent; single setal track between CU1 and CU2; FWL/HTL = 2.9 (2.7-3.1); FWL/FWW = 1.6 (1.6-1.7); FWFS/FWW = 0.05 (0.04-0.06); Max r-m to M/Min r-m to M = 2.3 (1.9-2.6); MV/PM = 0.9 (0.9-1.0); SV/MV = 1.2 (1.1-1.2); MV length/MV width = 2.4 (2.2-2.8). Hind wing width does not immediately decrease beyond hamuli; HWL/HWW = 7.3 (7.1-7.3); HWFS/HWW = 0.7.

Male

Antenna: Club segments loosely joined, C/F = 2.0 (1.8-2.0); F2/F1 = 1.3 (1.1-1.4); APB absent on funicle; 4-5 PLS on F1, 5-7 PLS on each of F2-C2, 4 PLS on C3; 6-7 BPS on each of F1-C1, 1 BPS on C2 and C3; 6-8 FS on F1, 9-12 FS on F2, 9-10 FS on C1, 9-10 FS on C2, 9-10 FS on C3, 8-9 FS on C4; US absent on F1-C3.

Genitalia: GL/GW = 4.0 (3.8-4.2); GL/HTL = 1.1; ADA/GL = 0.6; AI/GL = 0.1; PAR with terminal spine, subequal in width along entire length, their base anterior to posterior edge of ADA; PAR/GL = 0.5; VP base < half of capsule width, asymmetrically bifurcating in posterior third, with only one side of bifurcation approaching termination of capsule; VP/GL = 0.1; transverse hinge present; AP/GL =0.3; DR, VS absent.

Female

Unknown.

Other Material Examined. - **AUSTRALIA: Australian Capital Territory**: Wombat Crk., 6 km NE of Piccadilly Circus, 35°19'S, 148°51'E, 750m. el., ii.1984, Weir, Lawrence and Johnson, FIT (1♂). **Western Australia**: Stirling Range NP, 16.i.1987, J. Noyes (1♂) (QM).

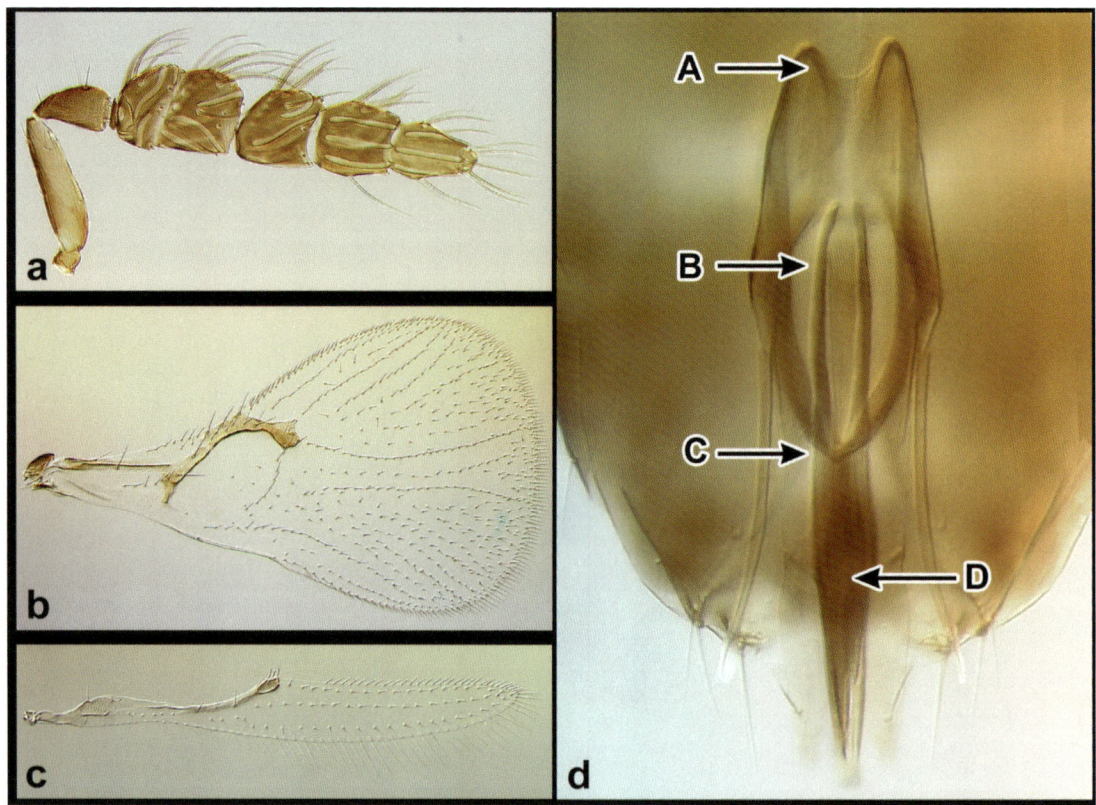

Figure 44. *Ufens placoides*, ♂. (a) antenna, lateral; (b) forewing, dorsal; (c) hind wing, dorsal; (d) genitalia, dorsal – arrows to {A} anterior invagination, {B} apodeme, {C} transverse hinge, {D} approximate location of asymmetric bifurcation of ventral process.

Ufens principalis Owen, 2005
(Fig. 45)
Ufens principalis Owen, in Al-Wahaibi et al., 2005: pp. 279-280

Diagnosis. - Forewing densely setose with broadly diverging setal tracks r-m to M; single setal track between CU1 and CU2. Hind wing width decreasing immediately apical of hamulus. Mescoscutal sculpturing longitudinally striate. Genitalia possessing parameres with terminal spine, subequal in width along entire length, their base even with posterior edge of anterodorsal aperture; dorsal ridge present; ventral process evenly tapering, width at base > half of capsule width.

Due to their very similar genitalia, the species most likely confused with *U. principalis* are *U. apollo, U. niger, U. similis,* and *U. taniae.* Of these, *U. apollo* is easily separable as it is the only member of this assemblage which has sparsely setose forewings. The absence of terminal parameral spines separates *U. taniae* from the others. The presence of a dorsal ridge also separates *U. principalis* from *U. apollo* and *U. taniae. U. principalis* can be separated from *U. similis* by the evenly tapering rather than laterally emarginate ventral process. Finally, *U. principalis* can be separated from *U. niger* by its longitudinally striate mesoscutal sculpturing, thicker marginal vein, genitalia with its anterior margin usually more transverse and posterior half less sinuate.

Types. - Holotype ♂, Allotype ♀ (USNM). **UNITED STATES: California:** *Riverside County*: Riverside, UCR Ag. Exp. Station, Field 7E, ex. *Homalodisca* sp. on jojoba, coll. 2.viii.2001, em. 8.viii.2001, Ali K. Al-Wahaibi. Paratypes 7♂, 16♀, same data; 2♂, 13♀ card mounted (1♂, 1♀ BMNH, remainder UCRC).

Distribution. - United States.

Biology. - *Ufens principalis* has been reared from *Homalodisca* species on host plants including *Simmondsia chinensis* (Link) C. K. Schneid. (Simmondsiaceae) , *Cercis* sp. (Fabaceae), *Parthenium argentatum* Gray (Asteraceae), *Baccharis salicifolia* (Ruiz, Lopez and Paron) (Asteraceae), *Schinus terebinthifolius* Raddi (Anacardiaceae), *Salix* nr. *lucida* (Salicaceae), *Vitis* sp. (Vitaceae), *Erythrina* sp. (Fabaceae), and several types of citrus (Rutaceae) (Al-Wahaibi et al. 2005).

Description. - BL 0.6 (0.6-0.7) mm. BL/HTL = 3.3 (3.1-3.6). Mesoscutal sculpturing longitudinally striate with interstitial sculpturing transverse. Forewing densely setose; AA present; single setal track between CU1 and CU2; FWL/HTL = 3.0 (2.9-3.1); FWL/FWW = 1.5 (1.4-1.6); FWFS/FWW = 0.06 (0.06-0.07); Max r-m to M/Min r-m to M = 3.8 (3.2-5.3); MV/PM = 1.2 (1.0-1.3); SV/MV = 0.9 (0.8-1.0); MV length/MV width = 3.0 (2.3-3.7). Hind wing width decreasing immediately apical of hamuli; HWL/HWW = 8.0 (7.8-8.3); HWFS/HWW = 0.9 (0.9-1.0).

Male

Antenna: Club segments loosely joined; C/F = 2.4 (2.2-2.5); F2/F1 = 1.0 (0.8-1.2); APB absent on funicle; 1 PLS on F1 and F2, 0-1 PLS on C1, 1 PLS on C2, 2 PLS on C3; 3-5 BPS on each of F1-C1, 1 BPS on C2 and C3; 9-12 FS on F1, 10-11 FS on F2, 9-11 FS on C1, 10-13 FS on C2, 8-10 FS on C3, 6-10 FS on C4; US absent on F1-C3.

Genitalia: Posterior half apparently not rigid; GL/GW = 3.6 (3.5-3.8); GL/HTL = 1.0 (0.9-1.1); ADA/GL = 0.4 (0.4-0.5); AI very shallow to absent, AI/GL = 0.004 (0-0.01); PAR with terminal spine, subequal in width along entire length, their base even with posterior edge of ADA; PAR/GL = 0.3 (0.3-0.4); VP evenly tapering, width at base > half of capsule width; VS/GL = ca. 0.2; DR extending ca. 0.1 or more of GL; AP, transverse hinge absent.

Female

Antenna: C/F = 2.0 (1.9-2.2); F2/F1 = 1.0 (0.7-1.4); 1 APB on F1 and F2, 1 APB on C3; 1 PLS on F1, 2 PLS on F2 and C1, 1-2 PLS on C2, 4 PLS on C3; 3-5 BPS on each of F1-C1, 1 BPS on C2 and C3; 0 FS on F1 and F2, 5-8 FS on C1, 10-13 FS on C2, 2-5 FS on C3; 1 UPP on C3; 10-14 US on F1 and F2, 9-17 US on C1, 0 US on C2, 4-5 US on C3.

Ovipositor: OL/HTL = 0.8 (0.8-1.0).

Other Material Examined. UNITED STATES: Arizona: *Cochise County*: Coronado National Forest: Barfoot Mtn., 11.ix.1978, G. Gordh (1♂); Dragoon Mtns., Jordan Canyon, 31°59.33'N, 110°01.07"W, 11.viii.2001, AKO, SW (1♂).

Santa Cruz County: Patagonia, 31°53'N, 110°77'W, 16.vi.1994, MT, E. Wilk and B. Brown (1♂). **California**: *Los Angeles County*: Altadena, 1.x/11.xii.1990; R. H. Crandell (1♂). *Riverside County*: *Riverside*: UCR Agricultural Operations, ex. *Homalodisca coagulata* in tangerine leaf, 13.v.1999, J. Bethke (1♂, 1♀); UCR Agricultural Operations, Field 7E, ex. *Homalodisca* sp. in jojoba, 3.vii.2000, A. K. Al-Wahaibi (1♂, 2♀); UCR campus, ex. *Homalodisca* sp. eggs in unknown leaf, 25.ix.2000, AKO, (4♂, 4♀; 2♂, 2♀ card mounted); UCR campus, ex. cicadellid eggs in *Erythrina*, 18.ix.2000, AKO, (1♂, 3♀; 2♀ card mounted); UCR Agricultural Operations, ex. *Homalodisca* sp. eggs in grapefruit leaf, 5.viii.2000, A. K. Al-Wahaibi (1♂); UCR Botanical Gardens, ex. *Homalodisca* eggs in Jojoba leaf, 30.viii.2001, AKO, (2♂, 1♀); UCR campus, ex. *Homalodisca* sp. eggs in redbud leaves, 03.x.2001, AKO and S. Triapitsyn (1♂, 1♀); UCR campus, ex. *Homalodisca* sp. eggs in redbud leaves, em. 2-3.x.2001, S. Triapitsyn (3♂, 3♀; 1♂, 1♀ card mounted); UCR Agricultural Operations, ex. *Homalodisca* sp. eggs on grape leaves, 19.viii.2002, R. Burks (1♂); Santa Rosa Plateau Ecological Reserve, 33°32.538'N, 117°14.758'W, MT, 14.viii - 7.ix.2001 (1♂). **New Mexico**: *Hidalgo County*: Gray Ranch, E slope Animas Mtns., Indian Creek Wash, N of Culberson Camp, 31°25.31'N, 108°40.52'W, SW, 5.viii.2002, JG and M. Gates (1♂).

Comments. - Volsellar length is difficult to measure as these structures are a challenge to distinguish in most slide-mounted specimens. Similarly, the extent of the dorsal ridge is often difficult to determine because in many specimens the darkened area indicating the ridge does not come to an abrupt termination, but instead gradually fades.

Molecular data for *U. principalis* as presented in Owen et al. (2007) can be found under Genbank accession numbers AY623530 (28S-D2+D3) and AY940372 (18S).

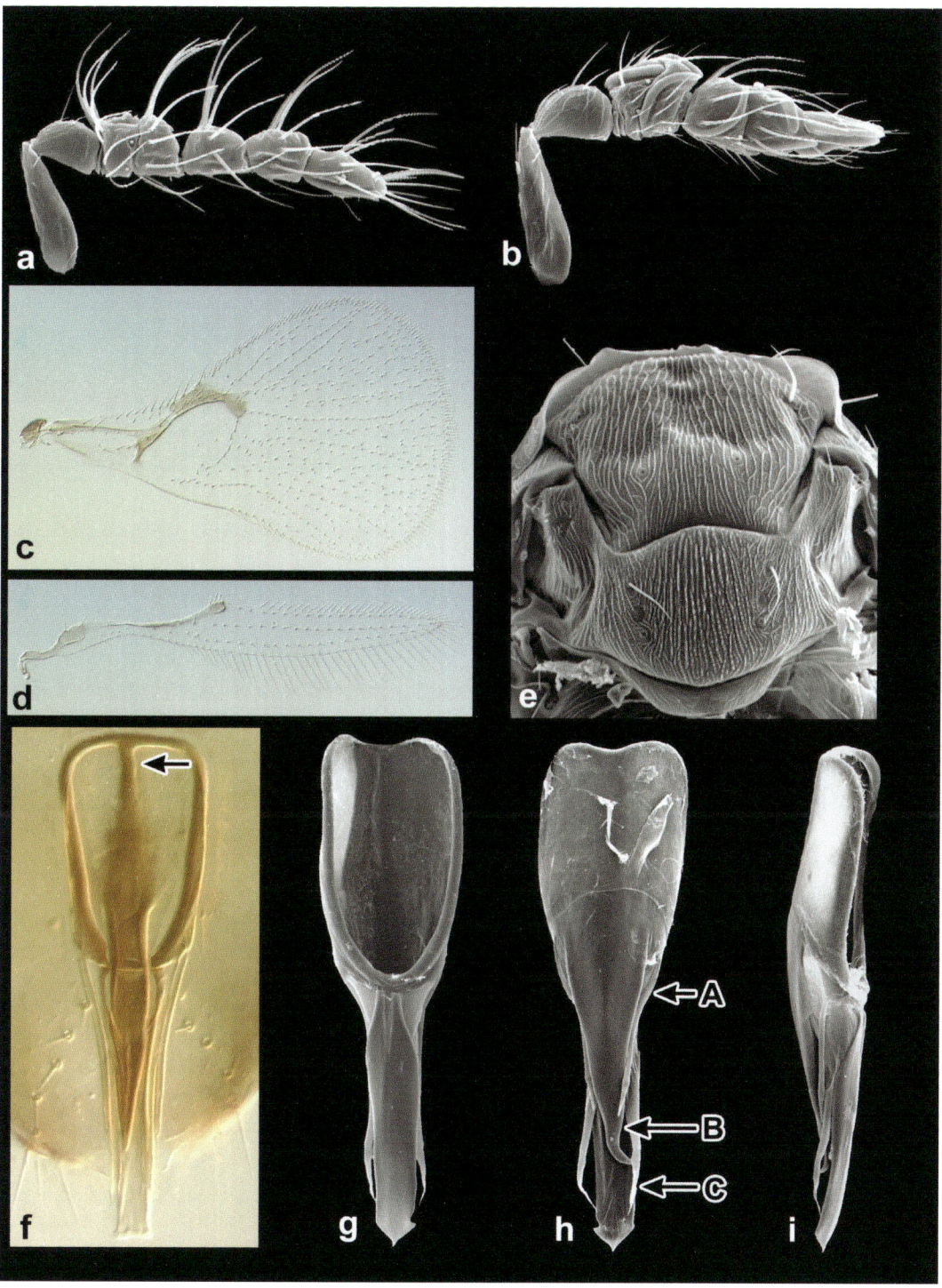

Figure 45. *Ufens principalis*. (a) ♂ antenna, lateral; (b) ♀ antenna, lateral; (c) forewing, dorsal; (d) hind wing, dorsal; (e) mesosoma, dorsal; (f) ♂ genitalia, dorsal – arrow to dorsal ridge; (g) ♂ genitalia, dorsal; (h) ♂ genitalia, ventral – arrows to {A} paramere base, {B} ventral process apex, {C} volsella apex; (i) ♂ genitalia, lateral.

Ufens rimatus **Lin, 1993**
(Fig. 46)

U. rimatus Lin, 1993: pp. 56-58.

Lin, 1994: pp. 211-212 (redescriptionand illustration).

Lin, 2002: pp. 349-350 (female described and host information given).

Diagnosis. - Antenna with flagelliform setae straight and bristle-like, and relatively few on terminal club segment. Forewing sparsely setose with narrowly diverging setal tracks r-m to M; single setal track between CU1 and CU2. Hind wing width decreasing immediately apical of hamuli. Mesoscutal sculpturing longitudinally striate. Genitalia with anterodorsal aperture distinctly narrower than capsule width; parameres short, subequal in width along entire length, with terminal spine, their base apical to posterior edge of anterodorsal aperture; volsellae straight and contiguous with lateral margin of capsule; dorsal ridge present; ventral process absent.

The flagelliform setae, which are relatively straight, bristle-like, and sparse on the last club segment, separate *U. rimatus* from all species other than *U. elimaeae*, to which it is most likely closely related. *U. rimatus*, however, is distinguished by the absence of unsocketed setae on the funicle, presence of volsellae, and parameres with a terminal spine. The relatively narrow anterodorsal aperture of *U. rimatus* is also shared with *U. invaginatus*, and *U. pallidus*. Other genitalic features of these species differ significantly, however (cf. Figs. 30, 41). In *U. pallidus*, the volsellae are crescent-shaped, not straight, and in *U. invaginatus* the basal margin of the capsule is deeply invaginated.

Types. - Holotype ♂ (FACS). **CHINA: Guangdong**: Guangzhou, Shipai, 3.xi.1985, N. Lin, SW. Paratypes 2♂ [1 examined], same data.

Distribution. - Borneo, China, Malaysia.

Biology. - Reared from the eggs of *Sophonia pallida* (Melichar) (Hemiptera) on guava, *Psidium guajava* L. (Myrtaceae).

Description (N=3). - BL 0.6 (0.5-0.6) mm. BL/HTL = 4.1 (3.9-4.2). Mesoscutal sculpturing longitudinally striate with interstitial sculpturing absent to rugulose. Forewing sparsely setose; AA present; single setal track between CU1 and CU2; FWL/HTL = 3.3 (3.1-3.4); FWL/FWW = 1.7; FWFS/FWW = 0.09 (0.07-0.1); Max r-m to M/Min r-m to M = 2.2 (1.9-2.4); MV/PM = 1.4 (1.3-1.5); SV/MV = 0.7 (0.7-0.8); MV length/MV width = 4.6 (3.9-5.3). Hind wing width decreasing immediately apical of hamuli; HWL/HWW = 8.2 (7.8-8.4); HWFS/HWW = 1.1 (1.0-1.3).

Male

Antenna: Club segments compact; C/F = 1.9 (1.7-2.0); F2/F1 = 1.3 (1.1-1.3); APB absent on funicle; 1 PLS on each of F1-C2, 2 PLS on C3; 3-4 BPS on each of F1-C1, 1 BPS on C2 and C3; FS rather straight and bristle-like, 3-5 FS on F1, 5-7 FS on F2, 7-8 FS on C1, 7-8 FS on C2, 5 FS on C3, 1-2 FS on C4; US absent on F1-C3. *Genitalia*: Capsule long, gradually tapering; GL/GW = 5.0 (4.3-5.7); GL/HTL = 1.5 (1.5-1.6); Width of ADA distinctly narrower than capsule width, ADA short, ADA/GL = 0.4; AI very shallow, AI/GL = 0.02 (0.01-0.02); PAR with terminal spine, subequal in width along entire length, their base posterior to posterior edge of ADA; PAR/GL = 0.2; VS contiguous with lateral margin of capsule, VS/GL = 0.4; DR extending ca. 0.05 of GL; AP, VP, transverse hinge absent.

Female (N=3)

Antenna: C/F = 2.0 (1.9-2.1); F2/F1 = 1.4 (1.1-1.8); 1 APB on F1 and F2, 0 APB on C3; 1 PLS on F1, 2-4 PLS on F2, 1-2 PLS on C1 and C2, 4 PLS on C3; 5-7 BPS on F1, 6 BPS on F2, 5-8 BPS on C1, 1 BPS on C2 and C3; 0 FS on F1 and F2, 4-5 FS on C1, 7-8 FS on C2, 3 FS on C3; 1 UPP on C3; 5-10 US on F1, 5-8 US on F2, 5-7 US on C1, 0 US on C2, 2-4 US on C3.

Ovipositor: OL/HTL = 2.0.

Material Examined. BORNEO: Sarawak: SW Gunung Buda, 64 km S Linmbang, 4°13'N, 114°56'E, 16-21.xi.1996, S. L. Heydon and S. Fung, MT (1♂, 2♀). **CHINA: Fujian**: Nanjing, 23.v.1991, N. Lin (1♀). **MALAYSIA: Sabah**: Danum Valley, 18.xii.1986-18.i.1987, M. Still (1♂).

Comments. - One of the male paratypes of this species was examined, and the bristle-like setae of the antenna and the distinctive male genitalia of this specimen are consistent with the original description of this species (Lin 1993).

The original description of this species designated a male holotype and two male paratypes (Lin 1993). In the subsequent description of the female (Lin 2002), two additional female specimens were examined, with the following locality information: China: Fujian: Fuzhou, 26°00'N, 119°23'E, N. Lin, Y. Chen and X. Fang, ex. eggs of *Sophinia pallida* (Melichar) on guava. An additional six males and two females were also apparently reared in this series (Lin 2005, pers. comm.). Although these specimens were not examined, it is assumed that they were correctly identified and represent a valid host record.

The female listed in material examined from Fujian was not collected in association with males and its identity is somewhat questionable. Although not incorporated into the description, this specimen is similar in all respects to associated females except for a slightly shorter ovipositor (OL/HTL = 1.5).

The coordinates of the type locality according to N. Lin (pers. comm.) are 23°00'N, 113°20'E.

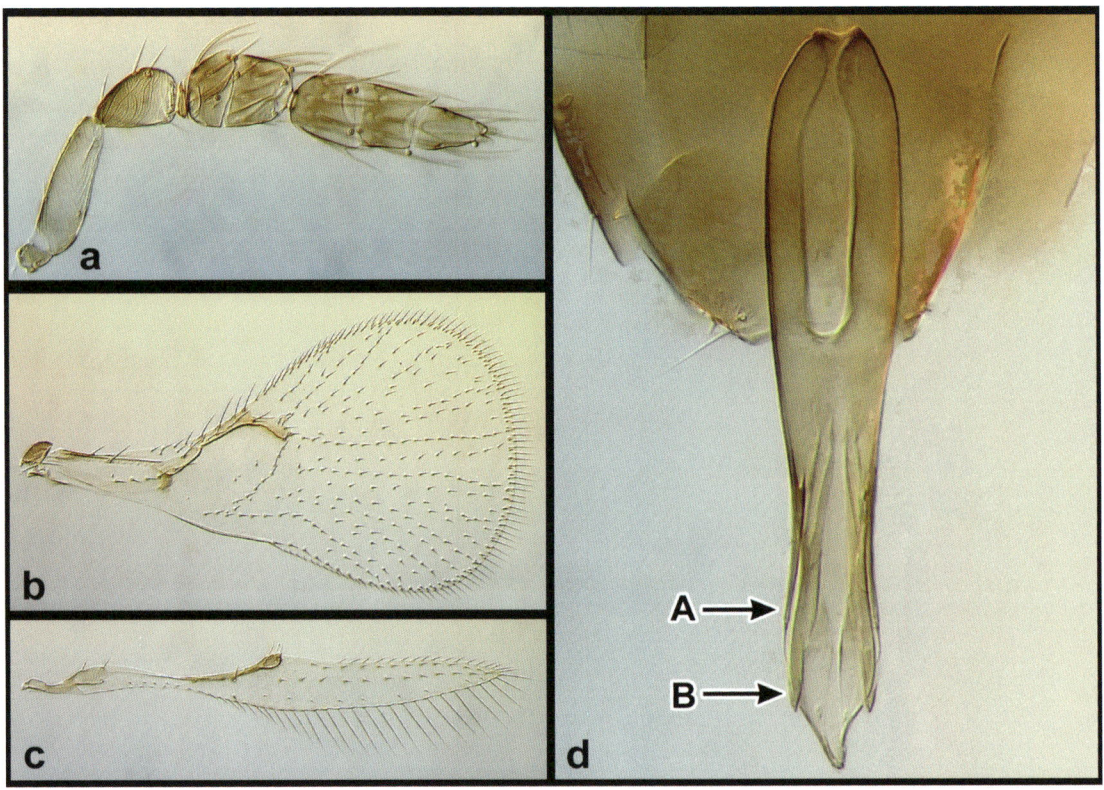

Figure 46. *Ufens rimatus*, ♂. (a) antenna, lateral; (b) forewing, dorsal; (c) hind wing, dorsal; (d) genitalia, dorsal – arrows to {A} paramere, {B} volsella. (Note that the dorsal ridge is not visible in figure, but is present.)

Ufens similis **(Kryger), 1932**
(Fig. 47)

Neocentrobia similis Kryger, 1932: pp. 42-43.
Ufens similis, Nowicki, 1935: pp. 573-574 (new combination).
Ufens similis megaloptila Nowicki, 1940: p. 624.
Viggiani, 1971: pp. 195, 203 (genitalia illustrated).
Lin, 1994: pp. 207-208 (redescription and illustration).
Ufens anomalus Lin, 1994: pp. 208-210, **new synonymy.**

Diagnosis. - Forewing densely setose with widely diverging setal tracks r-m to M and dispersed setae between CU1 and CU2. Hind wing width decreasing immediately apical of hamulus. Mesoscutal sculpturing longitudinally cellulate to striate. Genitalia possessing parameres with terminal spine, widest at base, and their base posterior to posterior edge of anterodorsal aperture; volsellae often laterally framing capsule; ventral process laterally emarginate, its width at base > half width of capsule.

      Ufens similis has genitalic and forewing characteristics very similar to *U. apollo*, *U. principalis*, *U. niger*, and *U. taniae*. It can be separated from all of these species by its laterally emarginate ventral process; in the other species this structure tapers evenly to the apex. *U. similis* can also be distinguished from *U. apollo* by its more densely setose forewing, and from *U. taniae* by its parameres with a terminal spine. It can be further separated from both *U. niger* and *U. principalis* by its narrower mesoscutal sculpturing, though there is intraspecific variation in both species. *U. similis* is similar to *U. niger* in that the genital capsule posterior of the anterodorsal aperture is somewhat sinuous in both.

Types. - Holotype ♀. **EGYPT**: Giza, 16.xi.1929, on *Convolvulus arvensis* L. Kryger (1932) lists the type to be in the collection of the Ministry of Agriculture, Giza.

U. anomalus. Holotype ♀, Allotype ♂ (FACS). **CHINA**: **Liaoning**: Shenyang, 27.vii.1992, N. Lin, SW. Paratypes 1♀ same data; 2♀, 1♂ same locality but 10.vii.1992; ♦1♀, Xiangyang, Beijing, 6.vii.1992, N. Lin, SW.

Distribution. - Asia, Europe, Africa.

Biology. - Unknown.

Description. - BL 0.5 (0.5-0.6) mm. BL/HTL = 3.2 (3.0-3.4). Mesoscutal sculpturing longitudinally cellulate to striate with interstitial sculpturing primarily longitudinal. Forewing densely setose; AA present; dispersed setae between CU1 and CU2; FWL/HTL = 3.1 (3.0-3.2); FWL/FWW = 1.8 (1.7-1.9); FWFS/FWW = 0.08 (0.07-0.1); Max r-m to M/Min r-m to M = 5.3 (4.3-7.0); MV/PM = 1.3 (1.2-1.5); SV/MV = 0.7 (0.6-0.7); MV length/MV width = 3.8 (3.6-4.1). Hind wing width decreasing immediately apical of hamuli; HWL/HWW = 9.7 (9.6-10.0); HWFS/HWW = 1.3 (1.2-1.4).

Male
Antenna: Club segments compact; C/F = 2.1 (1.9-2.3); F2/F1 = 1.2 (1.1-1.3); APB absent on funicle; 1 PLS on each of F1-C2, 2 PLS on C3; 2-5 BPS on each of F1-C1, 1 BPS on C2 and C3; 7-10 FS on F1, 8-13 FS on F2, 9-12 FS on C1, 10-13 FS on C2, 8-10 FS on C3, 6-8 FS on C4; US absent on F1-C3.

Genitalia: GL/GW = 2.7 (2.4-3.0); GL/HTL = 1.0 (0.9-1.2); ADA/GL = 0.6 (0.5-0.6); PAR with terminal spine, widest at base, their base posterior to posterior edge of ADA; PAR/GL = 0.3 (0.2-0.4); VS commonly conforming to lateral margins of aedeagus, often difficult to discern; VP laterally emarginate, its width at base > half of width of capsule; VP/GL= 0.5 (0.4-0.7); DR generally extending at least 0.2 of GL; AI, AP, transverse hinge absent.

Female (N=4)
Antenna: C/F = 2.2 (2.1-2.3); F2/F1 = 1.2 (1.1-1.4); 1 APB on F1 and F2, 1 APB on C3; 1 PLS on F1, 2 PLS on each of F2-C2, 4 PLS on C3; 2-5 BPS on F1, 4-5 BPS on F2, 3-4 BPS on C1, 1 BPS on C2 and C3; 0 FS on F1 and F2, 5-7 FS on C1, 7-9 FS on C2, 4-5 FS on C3; 1 UPP on C3; 7-12 US on F1, 9-14 US on F2, 10-13 US on C1, 0 US on C2, 3-4 US on C3.

Ovipositor: OL/HTL = 1.0 (0.9-1.2).

Material Examined. - **CHINA**: **Guangxi**: Nanning, 26.v.1986, Tang Yuqing (1♂, 1♀); Dongling, Shenyang, 9.vii.1992, N. Lin (1♀). **DENMARK**: 'Sortemose, Lillerod, 9.6.1930', 'J. P. Kryger prep.' (1♂). **FRANCE**: Dept. Hérault,

Rochelongue (nr. Agde), 20.vi.1979, J. T. Huber, SW riparian field (1♂); Dept. Gironde, Sainte Colombe (nr. Castillon-la-Bataille), 44°54'N, 00°02'W, MT, 2.vii.1998, M. van Helden (1♂). **GREECE**: **Corfu**: Agios Markos, 27.viii.1987, J. S. Noyes (1♀); Tebloni, 27.viii.1987, J. S. Noyes (1♂, 1♀). **INDIA**: **Uttar Pradesh**: New Delhi, IARI, 220m el., 28°37'51"N, 77°09'50"E, 8.ix.2003, JMH, YPT (1♂, 2♀); **Karnataka**: Mudigere, 994m. el., 13°07'09"N, 75°37'41"E, 26.ix.2003, JMH, SW grass and scrub (1♂). **ITALY**: **Lazio**: Roma Province, Castelporziano Presidential Estate, 41°41.95'N, 012°21.06'E, coastal dunes in N corner of estate, ~3m. el., Bologna, 12.vi.2003, JBM, AKO, JDP (1♂); **Sicily** (SR): Torre di Vendicari, 10km N Pachino, 4.vi.1992, JDP, S (1♂). **KENYA**: Laikipia District. Mpala Research Centre, Isecheno, 0°29'N, 36°90'E, 1650 m el., 17-30.ix.1999, R. Snelling, MT, savannah (1♂). **KYRGYZSTAN**: Dzhalal-Abad, Teke-Uyuk Ravine, 1850m el., 41°29'12"N, 74°35'50"E, vacuum, 30.vi.1999, C. H. Dietrich (2♂). **OMAN**: nr. old Muscat, grassy area in rocky gorge near water, SW, M. Huber and M. Reacher (1♂); Yiti, 23°31'N, 58°38'E, 20.iii.1992, cultivation, M. Gallagher (1♂). **SOUTH AFRICA**: **Gauteng**: Pretoria, iv.1957-iii.1958, D. Annecke, suction trap (2♂, 9♀); **Mpumalanga**: Nelspruit, Lowveld National Botanical Gardens, 25°29.53'S, 30°58.15'E, SW, 7.ii.2002, JG (1♂); **Western Cape**: The Baths, S of Citrusdale, 32°44.336'S, 19°02.346'E, 12.ii.2002, JG (1♀); Road to Treetops, ~5km S of The Baths along Olifant River, 32°47.399'S, 19°03.213'E, 14.ii.2002, SW, JG (1♂); Road to Treetops, ~4 km S of The Baths, 32°46.676'S, 019°03.167'E, fynbos, SW, JG (2♂). **THE GAMBIA**: Upper Baddibu District, E of Farafenni Dasilami, SW herbaceous plants at small stream, 8.xi.1992, M. Soderlund (1♀); Macarthy Isl. (S side), waterlogged meadow with bushes at river shore, 5.xi.1992, SW, M. Soderlund (1♂, 1♀).

Comments. - Although an anterior invagination is completely absent in most specimens of *U. similis*, in some there appears to be a very shallow invagination. The volsellae of *U. similis* can be very difficult to distinguish as they conform to the lateral margin of the capsule. They are generally only visible in specimens whose genitalia have been twisted or bent, allowing the volsellae to become separated. Even then, their base is indistinct, making them difficult to measure, though they likely extend from the posterior edge of the anterodorsal aperture to the apex of the genitalia.

Positive association of specimens with the type of *U. similis* is complicated by a number of factors. Firstly, the type material has not been examined, and in any case, is represented by only a single female. Secondly, the original description and drawings are rather poor (Kryger 1932). However, the densely setose forewing can be discerned from the original description, which separates it from the other Palearctic species *U. dilativena* and *U. foersteri*. Additionally, there is subsequent historical precedence for the recognition of *U. similis* with the characters herein outlined, as both Viggiani (1971) and Lin (1994) illustrated identical genitalia for this species.

Lin (1994) reports that *Ufens anomalus* was related to *U. similis*, but differentiated by "antennal club slender; gonobase of male genitalia wider, without chelate structures at the apex; aedeagus together with its apodemes about half as long as hind tibia." According to the drawing, the width of the antenna and capsule seem within the normal range of variation for *U. similis*. Although Lin reports that this species has apodemes, they are not labeled, and the separated 'aedeagus' from the drawing appears to simply be the posterior portion of the genital capsule. The 'chelate structures' that *U. anomalus* was reported to lack are presumably volsellae. It would not be surprising if volsellae were not identified. As indicated, these structures can be difficult to discern in *U. similis* as they are often closely appressed to the lateral margins of the aedeagus. In fact, it is likely that the thickened lateral margins of the posterior portion of the genitalia (the volsellae) are what Lin interpreted as apodemes. The species is therefore synonomyzed as it cannot be positively differentiated from *U. similis*.

Lin (1994) also reports *U. similis* from Poland, Egypt, and Turkey. Material from these locations was not located or examined.

Molecular data for *U. similis* was presented in Owen et al. (2007) as *Ufens* sp. 2, and can be found under Genbank accession numbers AY623534 (28S-D2+D3) and AY940376 (18S).

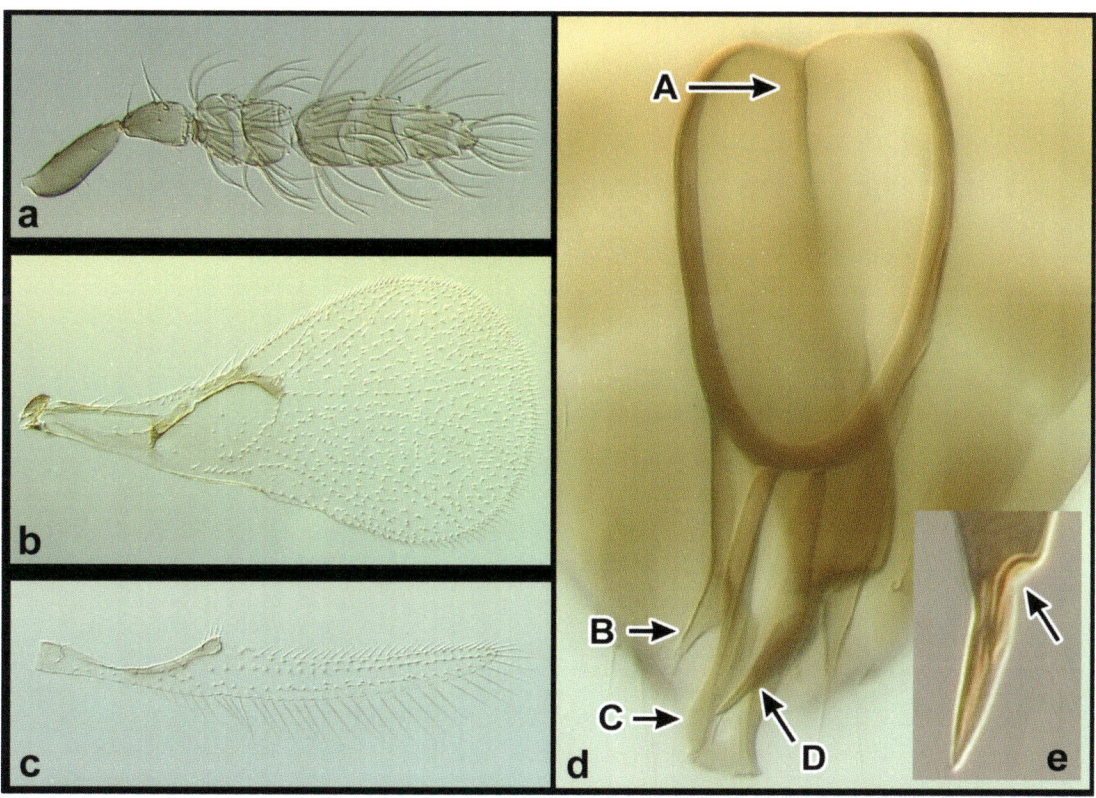

Figure 47. *Ufens similis*, ♂. (a) antenna, lateral; (b) forewing, dorsal; (c) hind wing, dorsal; (d) genitalia, dorsal – arrows to {A} dorsal ridge; {B} paramere, {C} volsella, {D} ventral process; (e) ventral process detail – arrow to lateral emargination.

Ufens simplipenis Owen, new species
(Fig. 48)

Diagnosis. - Forewing densely setose with widely diverging setal tracks r-m to M and dispersed setae between CU1 and CU2. Hind wing width decreasing immediately apical of hamuli. Mescoscutal sculpturing longitudinally cellulate. Genitalia highly simplified, consisting of a simple tube with minute apical parameres and volsellae; length less than (0.8-0.9x) hind tibial length; parameres with terminal spine, subequal in width along entire length, their base posterior to posterior edge of anterodorsal aperture; volsellae bifid apically.

Forewing characters such as the dense setation, broadly diverging setal track r-m to M, and dispersed setae between CU1 and CU2 suggest ties to *U. principalis*, *U. niger*, and *U. similis*. However, the simple tube-like genitalia with very small parameres and volsellae easily separate *U. simplipenis* from these species. In fact, the only species difficult to distinguish from *U. simplipenis* is *U. dolichopenis*. *U. simplipenis* and *U. dolichopenis* have the shortest parameres and volsellae relative to genitalia length known in the genus. However, *U. simplipenis* is distinguished by its considerably shorter male genitalia and ovipositor. (See comments of *U. dolichopenis* for a detailed discussion.)

Types. - ◆Holotype ♂, Allotype ♀ (USNM). **UNITED STATES: Oregon:** _Harney Co._: Fifteen Cent Lake, NE end, 10.vii.1999, JDP, SW.

Etymology. - Conjunction of the Latin simplis, or simple, and penis, in reference to the simple male genitalia.

Distribution. - Central and western North America.

Biology. - *U. simplipenis* has been reared from *Circulifer* (syn. *Eutettix*) *tenellus* (Baker)*, Aceratagallia abrupta* Oman, and *Homalodisca* sp. (all Hemiptera: Cicadellidae) from various localities in Baja California and the United States. It has also been reared from unidentified eggs on host plants including *Atriplex lentiformis* (Torr.) S.Watson. (Chenopodiaceae), grape (Vitaceae), *Chenopodium* sp. (Chenopodiaceae), *Lycium* sp. (Solanaceae), *Monolepis* sp. (Chenopodiaceae), *Plantago erecta* E. Morris (Plantaginaceae), *Pluchea sericea* (Nutt.) Coville (Asteraceae), *Salsola* sp. (Chenopodiaceae), *Sida hederacea* (Dougl.) Torr. (Malvaceae) and *Tamarix* sp. (Tamaricaceae).

Description. - BL 0.5 (0.5-0.6) mm. BL/HTL = 3.0 (2.8-3.2). Mesoscutal sculpturing longitudinally cellulate with interstitial sculpturing rugulose to longitudinal. Forewing densely setose, AA present, dispersed setae between CU1 and CU2; FWL/HTL = 3.0 (2.8-3.2); FWL/FWW = 1.7 (1.6-1.7); FWFS/FWW = 0.09 (0.08-0.1); Max r-m to M/Min r-m to M = 5.0 (3.4-6.6); MV/PM = 1.3 (1.2-1.5); SV/MV = 0.8 (0.6-0.9); MV length/MV width = 4.0 (3.4-4.9). Hind wing width decreasing immediately apical of hamuli; HWL/HWW = 8.8 (8.4-9.2); HWFS/HWW = 1.2 (1.2-1.3).

Male

Antenna: Club segments compact; C/F = 1.8 (1.0-2.0); F2/F1 = 1.2 (0.9-1.6); APB absent on funicle; 1 PLS on each of F1-C2, 2 PLS on C3; 2-3 BPS on each of F1-

C1, 1 BPS on C2 and C3; 10-15 FS on F1, 12-15 FS on F2, 11-15 FS on C1, 12-17 FS on C2, 9-13 FS on C3, 8-11 FS on C4; US absent on F1-C3.

Genitalia: Capsule laterally somewhat sigmoid; GL/GW = 4.7 (4.2-5.0); GL/HTL = 0.9 (0.8-0.9); ADA/GL = 0.5 (0.5-0.6); PAR with terminal spine, subequal in width along entire length, their base posterior to posterior edge of ADA; VS apically bifid; PAR and VS minute, apically placed and difficult to discern, ca. 0.1 of GL; AI, AP, DR, VP, transverse hinge absent.

Female

Antenna: C/F = 2.2 (2.0-2.4); F2/F1 = 1.2 (1.0-1.5); 1 APB on F1 and F2, 1 APB on C3; 1 PLS on F1, 2 PLS on F2, 2-3 PLS on C1, 2 PLS on C2, 4 PLS on C3; 2-4 BPS on each of F1-C1, 1 BPS on C2 and C3; 0 FS on F1 and F2, 5-8 FS on C1, 8-10 FS on C2, 4-5 FS on C3; 1 UPP on C3; 6-12 US on F1, 9-16 US on F2, 7-13 US on C1, 0 US on C2, 4-9 US on C3.

Ovipositor: OL/HTL = 1.0 (0.9-1.0).

Other Material Examined. - **CANADA: Alberta**: Medicine Hat, Kin Coulee, N Hwy. 1, 2190' el., 13.vi.1982, G. Gibson (1♂). **North Western Territories**: Martin River, FWI pipeline project, 1.viii.1972 (1♂). **MÉXICO: Baja California Sur**: Las Barracas, Santiago, ca. 30 km E, 21.iv – 2.vi.1989, P. Debach, YPT (1♂); Santo Domingo, 7.v.1928, C. F. Henderson, ex eggs of *Eutettix tenellus* (4♂, 2♀) [1 slide]. **Distrito Federal**: 12 mi. W Texcoco, 2300m el., 28.x.1982, J. T. Huber and A. Gonzalez (1♂). **Nuevo León**: Municipio Allende, Lazarillos de Abajo, 9.vii.1983, A. Gonzalez (1♂); Municipio Escobedo, Hda. El Canada, 12.vii.1983, G. Gordh (1♂, 1♀); 20 km N Salinas Victoria, Carretera 40, 13.vii.1983, A. Gonzalez H. (1♂); Santo Tomas, Bustamante, 18.v.1984, J. Sierra and M. Rodriguez (4♂); Terán, 8 km N, 16.v.1984, G. Gordh (1♂); Villa Damas, 18.v.1984, JDP, SW (1♂). **Sonora**: Caborca, 26.vii.1989 (2♂, 1♀). **Zacatecas**: Concepción del Oro, 4 mi. NE, 4.vii.1984, JBW (3♂, 1♀). **UNITED STATES: Arizona**: *Cochise Co.*: Chiricahua Mts., Sunnyflat, 29.vii.1979, J. LaSalle (1♂); Coronado NF, Texas Cyn., Texas Cyn. Rd., 32°02.20'N, 110°05.47'W, 11.viii.2001, AKO, SW (1♂); Coronado NF, Dragoon Mts., Jordan Cyn., 31°59.33'N, 110°01.07'W, 11.viii.2001, AKO, SW (1♂); Huachuca Mtns., 5200' el, 5364 Ash Cyn. Rd, 0.5 mi W Hwy 92, viii.1983, N. McFarland, MT (1♂, 3♀); 1 mi. NE Portal, 25.viii.1982, J. La Salle, SW (3♂, 1♀); 2 mi. E Bisbee, Jct. Hwy. 80 and Warren Rd., 27.viii.1982, J. La Salle, SW (1♀); Portal, 20 mi. N, 16.viii.1990, JDP, SW (1♂). *Coconino Co.*: Bitter Springs, Jct. Hwy. 89A/89, 6.ix.1997, M. Gates (2♂); Jacob Lake, 13 mi. S, 8800' el., 26.vi.1993, JDP, SW (1♂). *Graham Co.*: 2.4 mi. W on Hwy. 366 from Hwy 666, 1160m el., 27-28.v.1991, J. E. O'Hara, MT (1♂); 2.4 mi. W on Hwy. 366 from Hwy 191, 3800' el., 17-22.viii.1993, J. E. O'Hara, MT desert (2♂). *Maricopa Co.*: Gila River, nr. Gillespie Dam, 23.iii.1953, O. A. Hills, on *Monolepis* sp. (1♂, 2♀); 21 mi. S Gila Bend, 25.vii.1979, J. LaSalle (1♂); 3 mi E of exit 81, I-10 rest area, 27.iii.2003, M. Buffington, SW annuals (2♂). *Pima Co.*: Coronado NF, Box Cyn., 31°47.87'N, 110°46.49'W, 13.viii.2001, AKO, SW (1♂); Tucson, 26.vii.1961, G. D. Butler, suck sample from cotton (1♂); vicinity of Tucson, vii.1988, D. Gonzalez, ex. grape cuttings (1♂). *Pinal Co.*: Sacaton, 24.iii.1953, O. A. Hills, on *Monolepis* (5♂, 7♀) [1 slide]; Santa Catalina Mts., Peppersauce Campground, 24 km SE Oracle,

4700' el., 4.vi.1991, J. E. O'Hara, MT (1♂). _Santa Cruz Co._: Patagonia, 36°51'N, 110°77'W, 10-15.v / 16.vi.1994, E. Wilk and B. Brown, MT (4♂, 3♀); Nogales, N edge of town, 27.ix.1985, JDP, SW (1♂); Sycamore Canyon, 9 mi. W Pena Blanca, 4100' el., 12.viii.1983, Anderson (1♂); Coronado NP, Sycamore Canyon, 31°25.91'N, 111°11.31'W, 10.viii.2001, AKO, SW (1♂, 1♀); Coronado NP, Sycamore Canyon, 31°25.10'N, 111°11.31'W, 19.viii.2001, AKO, SW riparian (1♂). **California**: _Fresno Co._: Coalinga, 8.iv.1918, C. F. Henderson, ex. eggs of _Eutettix tenellus_ (1♂); Little Panoche, 7.iv.1928, C. F. Henderson, ex. eggs of _Eutettix tenellus_ (1♂); Big Panoche, 6.iv.1928, C. F. Henderson, ex. eggs of _Eutettix tenellus_ (1♂); Big Panoche, 19.iii.1938, W. Sottie, ex. eggs of _Eutettix tenellus_ (2♂); Coalinga Knolls [unverifiable locality], iv.1951, ex. _Circulifer tenellus_ (2♂, 2 ♀) [on 3 slides]. _Glenn Co._: Alder Springs and Hwy 162, 2.vi.1987, R. K. Velten, SW (1♂); 5 mi N Elk Creek, 5.vi.1987, JDP (1♂). _Imperial Co._: Seeley, Westside School Rd., 15.ii.1955, F. E. Skinner, ex. cage of _Chenopodium_ (2♂, 3♀); nr. Calexico, 21.ii.1956, F. E. Skinner (1♂); nr. Seeley, 21.ii.1956, F. E. Skinner (3♀); Calipatria, 16.iv-16.v.1959, R. Flock, on sugar beet ex _Circulifer_ (Multiple ♂ and ♀, on 7 slides); Niland, 8.v.1959, on _Lycium_ sp. (1♂); Seeley, 16.v.1959, R. A. Flock, on _Sida hederacea_, with _Circulifer tenellus_ and _Aceratagallia abrupta_ (4♂, 2♀) [1 slide]; Calipatria, 8.iii.1960, on _Chenopodium_ (1♂); Seeley, 20.iv.1960, R. C. Dickson (2♂, 5♀) [1 slide]; Finney Lake, nr. Brawley, 31.i.1982, on _Pluchea sericea_ (1 ♂); Imperial Valley, agricultural fields, iv-v.1991, YPT (1♂). _Inyo Co._: Amargosa Canyon, nr. Tecupa, W. Ewart, ex. _Pluchea_ (1♂); Big Pine, 2 mi E, 12.vii.1993, JDP, SW _Salix_, etc. along Owens River (2♂, 1♀); Goodale Crk., 4000' el., 36°59.10'N, 118°15.80'W, 14.ix.2001, AKO, SW desert and riparian (1♂, 1♀). _Kern Co._: Maricopa, 23.v-11.vi.1952, C. E. K., on russian thistle (_Salsola_ sp.) (1♂, 1♀); Edison, 25.v.1953, C. E. K., on low vegetation (3♂, 1♀) [1 slide]; Edison, 5.x.1953, Huffaker and Kennett, ex. thistle (3♂) [1 slide]; Gardner Field, 11.ix.1953, Huffaker and Kennett, ex. russian thistle (1♂); Oildale, 22.vii.1954, F. E. Skinner, ex. cage of russian thistle and fog weed (1♂); nr. Tupman, Elks Hills, 30.iii.1993, S. Triapitsyn, ex. sample of _Plantago erecta_ (1♂). _Lassen Co._: Eagle Lake (NE shore), Hwy. 139, ~3 mi. S Jct. A1, 24.vii.1992, JDP, SW _Salix_, nettle, etc. (1♂). _Los Angeles Co_: Altadena, 2.iv.1990-1.viii.1991, R. H. Crandall (4♂, 3♀). _Modoc Co._: 5.6 mi N Fandango Pass, 22.vii.1992, JDP, SW mesic meadow (_Salix_, etc.) (1♀). _Monterey Co._: Salinas Valley, ex. eggs of _Eutettix tenellus_ on _Chenopodium_ (1♂). _Nevada Co._: Sagehen Campground, 7km NW Hobart Mills, 30.vii.1994, S. L. Heydon (1♂). _Orange Co._: Irvine, San Joaquin Freshwater Marsh Res., 11.vi.1986, J. LaSalle, SW (1♂, 1♀). _Riverside Co._: Blythe, 21.i.1960, R. C. Dickson, on _Atriplex lentiformis_ (2♂, 1♀) [on 2 slides]; Coachella (Ave. 62), 1.v.1986, W. White, ex. leafhopper (_Homalodisca_?) eggs on _Tamarix_ (4♂, 1♀); Thermal, Grant and Ave. 62, 7.v.1986, W. White and M. Moratorio, ex. _Homalodisca_ eggs on _Tamarix_ #5 (2♂, 1♀); Lake Mathews, S end of (S of Cajalco Rd.), 13-16.iv.1993, G. Bruyea and JDP, YPT assoc. w/ _Encelia_ (1♂); Lake Skinner, NE, (MET B11) burned, ca 1580' el., 33°36'01"N, 117°01'58"W, 15-

29.viii.1996 / 16-27.iii.1997 / 24.iv-8.v.1997, JDP, MT coastal sage scrub (3♂, 2♀); as above, N, (MET B4), 30.vii-13.viii.1998 (1♂, 2♀); as above, N, (MET U4) unburned, 33°36'04"N, 117°02'18"W, 15-26.iii.1998 / 16-30.vii.1998 (3♂, 1♀); as above, NE, (MET U11) unburned (1♂); Menifee Valley, Hills on W end, 1800' el., 28.vi.1983, JDP, SW ravine bottom (1♂); Menifee Valley, Hills on W end, 1800' el., 33°19'N, 117°13'W, 12.vii.1995 / 31.vii.1995 / 1.viii.1995 / 1-29.ii.1996, JDP, MT (5♂, 4♀); Mission Creek Rd., ca. 8 mi. W Desert Hot Springs, 16.v.1985, J. T. Huber, SW marsh and desert vegetation (2♂, 1♀); Pacific Crest Trail N of Hwy. 74, 4900' el., 33°33.81'N, 116°34.62'W, 17.v.2001, JDP, SW (2♂, 2♀); Santa Rosa Plateau Ecol. Reserve, 590m. el., 33°32.52'N, 117°14.64'W, 7-28.iv.2002, JDP, MT #2 (2♂). *San Bernardino Co.*: I-15, 5.6 mi SW Baker, Zyzyxx Rd. exit, 30.iii.1989, JDP, SW *Larrea, Bebbia*, etc. (1♂); Clark Mts (W end), ca. 1490m el., 35°31'45"N, 115°38'15"W, 23.v.2001, JDP, SW (1♂); Granite Mts. Reserve, Granite Cove, 34°48'N, 115°39'W, v.14-17.1994, JDP and GP, SW (2♂, 2♀); Granite Mts. Reserve, Granite Cove, 34°48'N, 115°39'W, 26.viii.1994, GP and A. Urena, D-Vac desert vegetation *Acacia greggii*, etc. (2♂, 1♀); San Gorgonio Wilderness, 1 mi N Aspen Grove, 19.viii.1982, J. T. Huber, SW (1♂); San Gorgonio Wilderness, Fish Creek Trail, 8600' el., 19.vii.1982, JBW (1♂, 1♀); Twentynine Palms, Mesquite Springs, 12.iv.1984, J. LaSalle, SW (1♂, 1♀); Summit Valley, 1.7 mi E of I-15, 28.v.1981, J. LaSalle (1♂); San Bernardino Mts, Holcomb Valley, 6000' el., 23.vi.1982, J.T. Huber, SW (1♀); Holcomb Valley Rd. and Van Dusen Cyn. Rd., 16.vi.1988, R. K. Velten, SW *Ceanothus*, etc. (1♀); San Bernardino N. F., 1 mi. N Big Bear City, Van Dusen Cyn. Rd., 10.vii.1988, R. K. Velten, SW *Salix*, etc. (1♂); same location, 23.vi.1989, R. K. Velten, SW *Salix*, etc. (1♂, 1♀); same location, 34°16.930'N, 116°51.644'W, 10.v.2002, AKO, YPT riparian wash and scrub (1♂). *San Diego Co.*: Anza-Borrego State Park, Coyote Canyon, 1 mi W Ocotillo Flat, 14.v.1991, JDP, SW riparian habitat (1♂). *San Luis Obispo Co.*: 6 mi SE Pozo, 1500' el., 10-25.iii.1990, W. E. Wahl, MT (1♂). *Shasta Co.*: 10 km N Lakehead, 6.ix.1995, L. A. Baptiste, SW *Solidago* sp. (2♂). *Siskiyou Co.*: Bartle, ca. 2 mi. W along McCloud River, 17.vii.1990, JDP, SW *Salix*, etc. (1♀). *Solano Co.*: Cold Canyon Reserve, 11 km W. Winters, 11.x.1997, S. L. Heydon, off *Baccharis pilularis* (1♂). *Tulare Co.*: Lindcove Field Station, Lambs quarter, 14.iv.1984, R. Milner (6♂, 8♀). *Ventura Co.*: Camarillo, 26.viii.1953, C. E. K., on thistle (1♂); Camarillo, 25.ix.1953, C. E. K., on saltbush (2♂) [1 slide]; Lake Piru, 350m, 16.ii.1996, M. Gates (1♂). **Colorado**: *Adams Co.*: Brighton, 5.viii.1992, S. L. Heydon (1♂). *Teller Co.*: Woodland Park, 7 mi. N, South Meadows Camp, 21-28.vii.1977, S. and J. Peck (1♂). **Illinois**: *Champaign Co.*: Urbana, 3.ix.1983, J. T. and D. E. Huber, SW (1♂, 3♀). *Mason Co.*: Sandridge State Forest, 6.vii.1980, S. Heydon (2♂, 2♀). *Washington Co.*: Dubois, 4.ix.1983, J. and D. Huber, SW clover, grasses, swamp vegetation (1♂). **Missouri**: *Boone Co.*: Columbia, Hinkston Creek, 8.ix.1987, JDP, SW (1♂, 1♀). **Montana:** *Silver Bow Co.*: Butte, 23.vii.1983, JDP, SW riparian (1♂). **Nevada**: *Elko Co.*: Carlin, 11.vii.1985, JDP, SW along Humboldt River (1♂). **New Mexico**: *Hidalgo Co.*: Gray Ranch, Cienega, S of main office, 31°31.72'N, 108°52.83'W, 6.viii.2002, JG and M. Gates, SW (1♂, 1♀); Gray Ranch, ca. 0.5 mi. S headquarters, 31°31.72'N,

108°52.82'W, swampy impoundments, 6.viii.2002, JG and M. Gates, (1♂); Lordsburg, 15 mi. NE on Hwy. 90, 25.viii.1982, J. LaSalle, SW (2♂, 1♀). *Valencia Co.*: Las Lunas, 20 mi. W, Carrizo Arroyo, 1-23.viii.1977, S. and J. Peck, MT along streambed (3♂). *Quay Co.*: Tecumcari, along Rt. 66, 4.vi.2003, M. Buffington, SW vegetation in city (2♂, 1♀). **Oregon**: *Baker Co.*: nr. Unity 24.iv.1982, G. Gordh (1♂). *Clackamas Co.*: Zig Zag, 22.vii.1988, JDP, SW riparian vegetation (2♀). *Harney Co.*: Alvord Desert Rd., 2.5 mi. SW Hwy. 78, 10.vii.1999, JDP and D. G. Pinto, YPT (1♂); Mann Lake, NE end, 10.vii.1999, JDP, SW (1♂, 1♀); Steens Mtn., Loop Rd. at Blitzen Crossing, 9.vii.1999, JDP, SW (1♂, 1♀). *Joseph Co.*: 15 mi. W. Glendale, 9.vi.1985, P. Hanson (1♂). *Lake Co.*: Valley Falls, 5.4 mi. S, 5.viii.1995, JDP, SW pine, juniper, willow (1♂); Valley Falls, 11 mi. NW, 5.viii.1995, JDP, SW *Chrysothamnus*, *Sarcobatus*, etc. (1♂, 1♀). *Malheur Co.*: 4.5 mi. W of Jordan Valley, 11.vii.1999, JDP, SW *Salix*, etc. (1♂, 1♀). **Texas**: *Brewster Co.*: Big Bend NP, Cottonwood Campsite, 2300' el., 13-14.vii.1982, G. A. P. Gibson (2♂); Big Bend NP, 20.iii.1992, JBW and R. Wharton, SW (1♂); Big Bend NP, "No. Rosillos Mts." [N of Rosillos Mts.?], 29°34'N, 103°15'W, 17-21.iii.1992, JBW and R. Wharton (4♂, 1♀); Big Bend NP, No. Rosillos Mts., 4.x.1991, JBW, SW (1♂); Big Bend NP, Rosillos Mts., Buttrill Spring, 23.iv.1991, G. Zolnerowich (2♂, 1♀). *Dimmit Co.*: Chaparral Wildlife area, pasture 15-E Guajalote, 30.ix.1990, JBW (1♂). *Jim Wells Co.*: 8 mi. W of Ben Bolt, La Copita Research Station, 20.v.1987, JBW (2♂, 1♀); as above, area near pond, 29.ix.1990, JBW, SW (1♂); as above, 28-30.ix.1990, R. Wharton (1♂); as above, 0.5 mi. S tank, 24.iii.1990, G. Zolnerowich (1♂); as above, North Fence Pasture 52, 23.iii.1990, G. Zolnerowich (1♂). *Kerr Co.*: Center Point, 31.vi – 6.viii.1987, R. Wharton (1♂). *Presidio Co.*: Big Bend Ranch SNA, 2.5 mi. W of La Saucedo, 9.viii.1991, JBW, SW (1♂); 2.8 mi. E of La Saucedo of Big Bend Ranch SNA, 27-28.iv.1991, G. Zolnerowich, YPT (1♂, 1♀); Big Bend Ranch SNA, 3.5 mi. W of La Saucedo, 26-28.iv.1991, G. Zolnerowich, YPT (1♀); Big Bend Ranch SNA, McGuirk's Tank, 19.vi.1991, JBW, SW (♂). *Ward Co.*: 1 mi. S of Grandfalls, 19.iv.1985, J. C. Shaffner (3♂, 2♀). **Utah**: *Garfield Co.*: 7.2 mi. S of Ticaboo, Cane Springs Desert, 20.v.1995, JDP, SW desert flowers (1♂). *San Juan Co.*: Abajo Mtns., 8600' el., 4.2 mi. SE Indian Creek, 27.vi.1993, JDP, SW (1♂). *Washington Co.*: 14 mi. SW of Shivwits, 19.iv.1994, JDP, SW desert vegetation (3♂, 1♀). *Wayne Co.*: 6 mi. W of Caineville, along Fremont River, ca. 4700' el., JDP, SW (3♂, 3♀); 15.6 mi. N of Hanksville, 20.v.1995, JDP, SW desert flowers (1♂). **Wisconsin**: *Milwaukee Co.*: Milwaukee, Fox Point Suburb, 2.ix.1983, J. T. Huber, SW (1♂, 1♀). **Wyoming**: *Carbon Co.*: 17 mi. E of Rawlings, 1.5 mi. N I-80 at North Platte River, 7.vii-2.ix.1991, S. Shaw, MT (1♂, 1♀).

Comments. - *U. simplipenis* is one of the most commonly collected species in the Nearctic, especially in the western United States. In most slide-mounted specimens, the parameres and volsellae are very difficult to discern due to their small size and their lateral placement. This makes them difficult to distinguish from the capsule margin. Nevertheless, the parameres especially are visible in some slide-mounted specimens. The presence of parameres and volsellae were verified with SEM (Fig. 48 j and k).

Molecular data for *U. simplipenis* was presented in Owen et al. (2007) as *Ufens* sp. 10, and can be found under Genbank accession numbers AY623539 (28S-D2+D3) and AY940380 (18S).

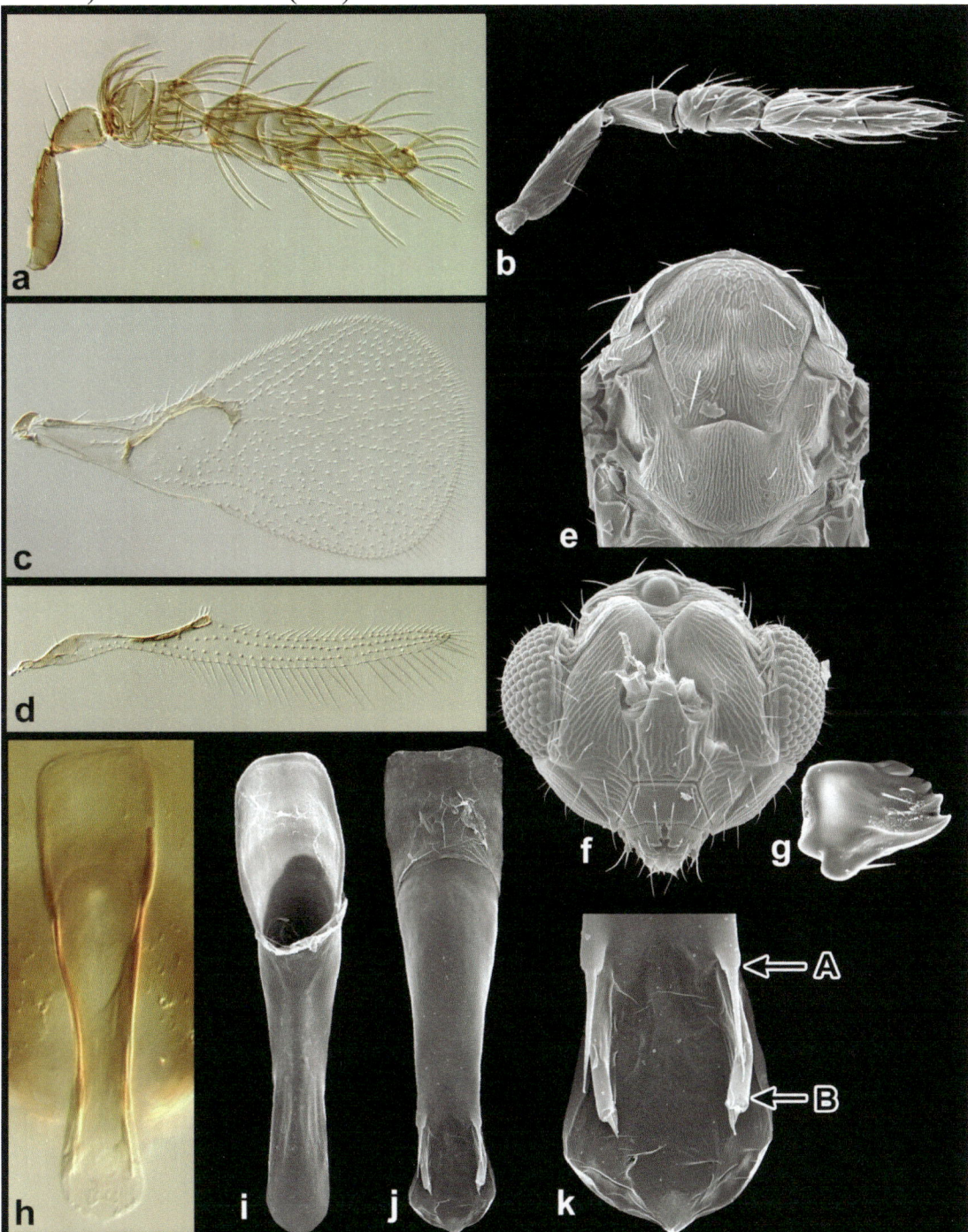

Figure 48. *Ufens simplipenis*. (a) ♂ antenna, lateral; (b) ♀ antenna, lateral; (c) forewing, dorsal; (d) hind wing, dorsal; (e) mesosoma, dorsal; (f) ♂ head, anterior; (g) mandible, posterior; (h, i) ♂ genitalia, dorsal; (j) ♂ genitalia, ventral; (k) ♂ genitalia, ventral detail of apex – arrows to {A} paramere, {B} volsella.

Ufens spicifer Owen, new species
(Fig. 49)

Diagnosis. - Forewing sparsely setose with narrowly diverging setal tracks r-m to M and an incomplete setal track between CU1 and CU2. Hind wing width decreasing immediately apical of hamulus. Mesoscutal sculpturing longitudinally striate. Genitalia capsule narrow; parameres small and with terminal spine, their width subequal along entire length, their base posterior to posterior edge of anterodorsal aperture; volsellae rigid and spine-like; no other appendages present.

Both the antenna and the genitalia of this species are distinctive. The extremely small terminal club segment is a trait shared only by *U. gloriosus* and *U. messapus*. Both of these species also share the presence of APB on the funicle and lack flagelliform setae on F1. However, unlike *U. spicifer*, *U. gloriosus* does not have US on the funicle or club, and both species have flagelliform setae on F2. In terms of genitalia, neither *U. gloriosus* or *U. messapus* are likely to be confused with *U. spicifer* as neither have a similiarly narrow genital capsule nor its rigidly straight volsellae. The only species other than *U. spicifer* with simplified and uniformly narrow genitalia are the North American *U. dolichopenis* and *U. simplipenis*, though the volsellae in these two species are minute and bifid (at least certainly for *U. simplipenis*, as verified by SEM).

Types. - ♦Holotype ♂ (ANIC). **AUSTRALIA: Queensland**: Cockatoo Ck. Xing, 17km NW Heathlands, 11°39'S, 142°27'E, 26.i-29.ii.1992, P. Feehney, MT #5 in open forest.

Etymology. - Latin for bearing a spike, in reference to the spike-like volsellae.

Distribution. - Australia.

Biology. - Unknown.

Description. - BL 0.6 mm. BL/HTL = 3.8. Mesoscutal sculpturing longitudinally striate with interstitial sculpturing light and primarily longitudinal. Forewing sparsely setose; AA absent; single, incomplete setal track between CU1 and CU2; FWL/HTL = 3.0; FWL/FWW = 1.7; FWFS/FWW = 0.1; Max r-m to M/Min r-m to M = 1.2; MV/PM = 1.1; SV/MV = 1.0; MV length/MV width = 3.1. Hind wing width decreasing immediately apical of hamuli; Hind wing broad, HWL/HWW = 4.1; HWFS/HWW = 1.1.

Male (N=1)

Antenna: Club segments very compact, C4 minute and not extending beyond PLS of C3; club comparatively long, C/F = 2.8; F2/F1 = 0.9; 1 APB on F1 and F2; 1 PLS on each of F1-C2, 2 PLS on C3; 3 BPS on each of F1-C1, 1 BPS on C2 and C3; 0 FS on F1 and F2, 5 FS on C1, 6 FS on C2, 5 FS on C3, 2 FS on C4; 5 US on F1, 7 US on F2, 5 US on C1.

Genitalia: Capsule long, narrow and mainly parallel-sided; GL/GW = 6.8; GL/HTL = 1.5; ADA/GL = 0.6; PAR small, with terminal spine, subequal in width along entire length, their base posterior to posterior edge of ADA; PAR/GL = 0.2; VS long, rigid and evenly tapering to an apical point, VS/GL = 0.3; AI, AP, DR, VP, transverse hinge absent.

Female

Unknown.
Other Material Examined. - None.

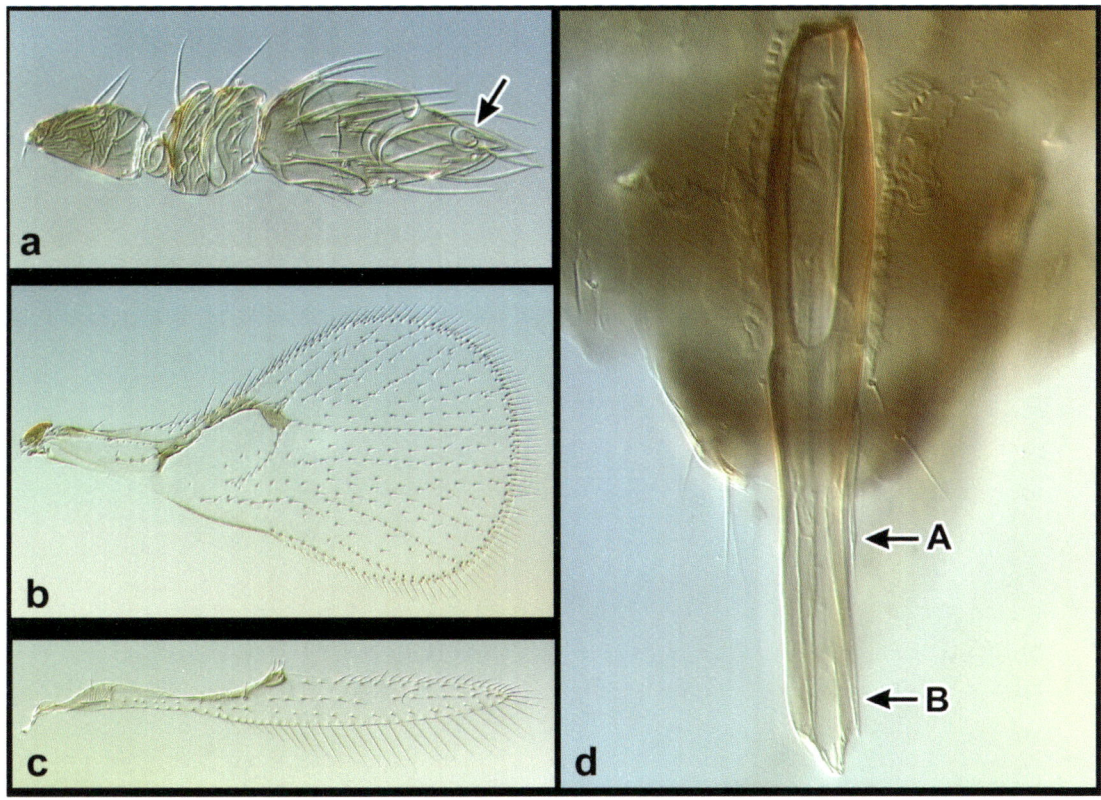

Figure 49. *Ufens spicifer*, ♂. (a) antenna, lateral – arrow to minute terminal club segment; (b) forewing, dorsal; (c) hind wing, dorsal; (d) genitalia, dorsal – arrows to {A} paramere, {B} volsella.

Ufens taniae **Owen, new species**
(Fig. 50)

Diagnosis. Forewing densely setose with moderately diverging setal tracks r-m to M and a single setal track between CU1 and CU2. Hind wing width decreasing immediately apical of hamuli. Mesoscutal sculpturing longitudinally cellulate to striate. Genitalia possessing parameres without terminal spine, widest and spatulate apically, their base even with posterior edge of anterodorsal aperture; volsellae filiform but difficult to distinguish; ventral process symmetrical, its basal width greater than half of capsule width.

　　U. taniae has genitalic and forewing characteristics very similar to *U. apollo*, *U. principalis*, *U. niger*, and *U. similis*. *U. taniae* is readily separated from these species by its apically spatulate parameres, which lack a terminal spine. It is further separated from *U. principalis*, *U. niger*, and *U. similis* by its forewings, which have a single setal track between CU1 and CU2, rather than dispersed setae. *U. taniae* does not have the laterally emarginate ventral process found in *U. similis*, nor the cellulate sculpturing occurring in *U. niger*.

Types. ♦Holotype ♂ (USNM). **COSTA RICA: Provincia Puntarenas**: "P. Int. La Amistad," Estation Altamira, sendero a Casa Coco, ii.2002, 1700m, C. Hanson and parataxónomos, MT, "L_S_331750_574400, #67021". Paratype 1♂, same data (UCRC).

Etymology. - Named for my wife, Tania Kim.

Distribution. - Central and South America (Argentina).

Biology. - Unknown.

Description (N=3). - BL 0.6 (0.5-0.7) mm. BL/HTL = 3.4 (3.3-3.6). Mesoscutal sculpturing longitudinally cellulate to striate with interstitial sculpturing transverse. Forewing densely setose; AA present; single setal track between CU1 and CU2; FWL/HTL = 3.3 (3.2-3.3); FWL/FWW = 1.6 (1.5-1.7); FWFS/FWW = 0.08 (0.07-0.09); Max r-m to M/Min r-m to M = 3.0 (2.7-3.4); MV/PM = 1.2 (1.1-1.2); SV/MV = 0.9; MV length/MV width = 3.2 (2.8-3.4). Hind wing width decreasing immediately apical of hamuli; HWL/HWW = 10.0 (9.1-10.6); HWFS/HWW = 1.2 (1.1-1.3).

Male

Antenna: C/F = 2.4 (2.3-2.5); F2/F1 = 1.3 (1.0-1.6); APB absent on funicle; 1 PLS on each of F1-C2, 1-2 PLS on C3; 2-4 BPS on each of F1-C1, 1 BPS on C2, 1-2 BPS on C3; 5-10 FS on F1, 9-10 FS on F2, 7-8 FS on C1, 8-10 FS on C2, 6-9 FS on C3, 5-7 FS on C4; US absent on F1-C3.

Genitalia: GL/GW = 2.8 (2.5-3.1); GL/HTL = 0.9 (0.8-1.0); ADA/GL = 0.5; PAR without terminal spine and spatulate apically, widest near apex, their base even with posterior edge of ADA; PAR/GL = 0.4 (0.4-0.5); VS filiform, difficult to discern; VP symmetrical, its width at base > half of width of capsule; VP/GL= 0.5 (0.4-0.5); DR absent or obsolescent; AI, AP, transverse hinge absent.

Female

Antenna: C/F = 2.0; F2/F1 = 1.0; 1 APB on F1 and F2, 1 APB on C3; 1 PLS on F1, 2 PLS on each of F2-C2, 4 PLS on C3; 3-5 BPS on each of F1-C1, 1 BPS on C2 and C3; 0 FS on F1 and F2, 4-5 FS on C1, 6-7 FS on C2, 3 FS on C3; 1 UPP on C3; 8-9 US on F1, 7-12 US on F2, 9-12 US on C1, 0 US on C2, 2-3 US on C3.

Ovipositor: OL/HTL = 1.0 (0.9-1.0).

Other Material Examined. - **COSTA RICA**: 2♀ with same data as types. **ARGENTINA: La Rioja:** Chuquis, 1575m el., 28°53'40"S, 67°00'31"W, 17.iii.2003, JBM, SW *Acacia* scrub (1♂) (USNM).

Comments. - As in *U. similis*, the volsellae of *U. taniae* are difficult to discern, being clearly visible only in the paratype from Costa Rica. Presumably, in the other specimens the volsellae are closely appressed to the posterior portion of the genitalia. SEM verification is needed.

The specimen from Argentina is the only *Ufens* known from South America. Interestingly, this argentine specimen has a somewhat different ventral process than the other specimens examined. Its ventral process is gradually tapering, whereas that of the Costa Rican material is wider throughout its length and then abruptly tapering near its apex. No other character suggests a partitioning of this species, though further evaluation with additional specimens would be useful.

Molecular data for *U. taniae* was presented in Owen et al. (2007) as *Ufens* sp. 3, and can be found under Genbank accession numbers AY623542 (28S-D2+D3) and AY940383 (18S).

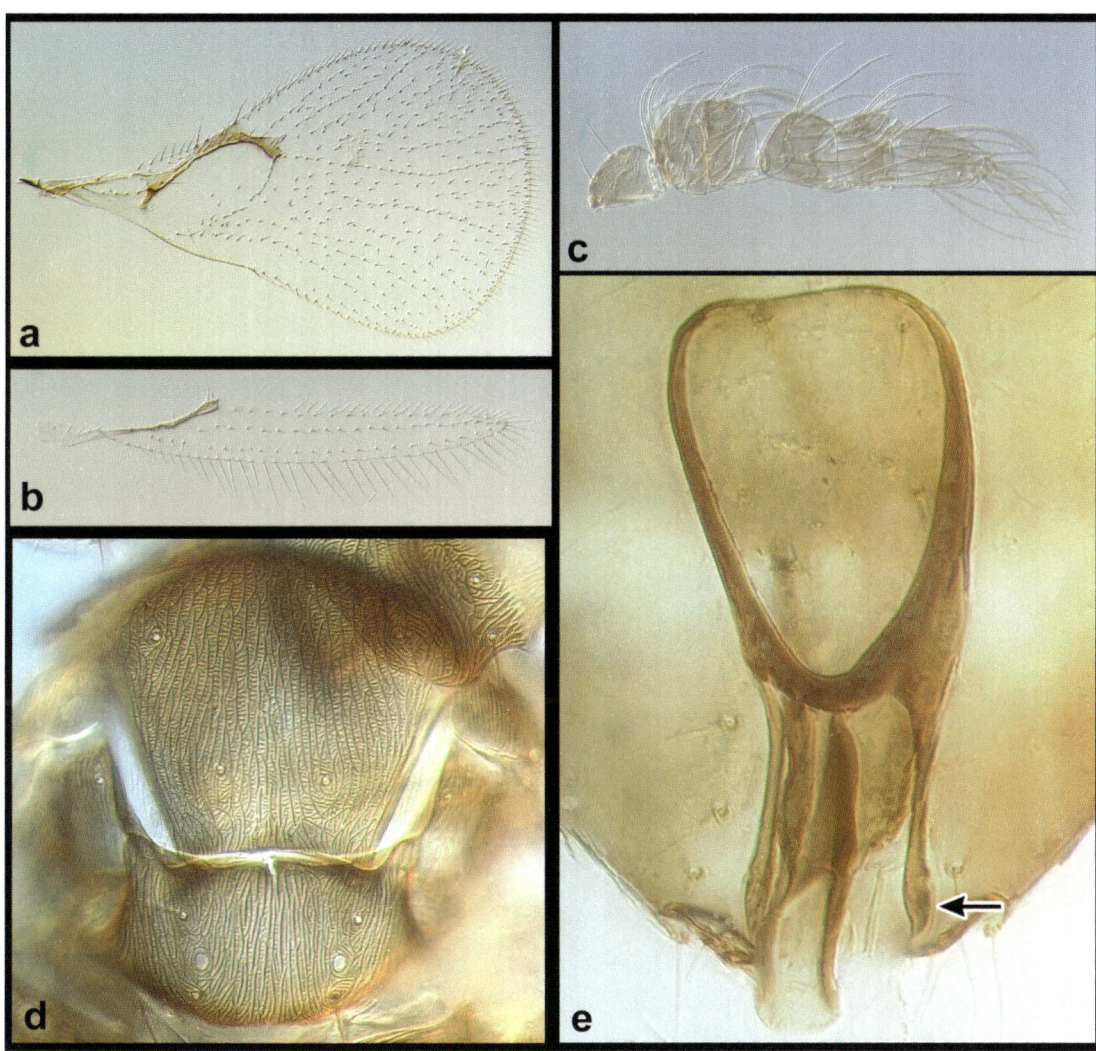

Figure 50. *Ufens taniae*, ♂. (a) forewing, dorsal; (b) hind wing, dorsal; (c) antenna, lateral; (d) mesosoma, dorsal; (e) genitalia, dorsal – arrow to spatulate apex of paramere.

Ufens thylacinus Owen, new species
(Fig. 51)

Diagnosis. - Forewing sparsely setose with narrowly diverging setal tracks r-m to M and a single setal track between CU1 and CU2. Hind wing width not decreasing immediately apical of hamuli. Mesoscutal sculpturing longitudinally cellulate to striate. Genitalia capsule narrow; ventral process long, apparently hollow at base; transverse hinge present; small paired appendages at apex; no other appendages present.

The only other species without volsellae and parameres but with a ventral process is *U. flavipes*. *Ufens thylacinus* can be differentiated from this species by its transverse hinge which is immediately posterior of the anterodorsal aperture, ventral process with an apparently hollow base, abrupt constriction in the posterior half of the anterodorsal aperture, and distinct paired apical appendages. The genitalia of *U. thylacinus* bear the most resemblance to those of *U. austini*. It can be differentiated from that species by the lack of parameres, as well as the anterior position of the transverse hinge and the paired apical appendages.

Types. - ♦Holotype ♂ (QM). **AUSTRALIA: Queensland**: Mt. Isa, 20 km ENE, 20°40'S, 139°41'E, 3-4.iii.2002, 370m el., C. J. Burwell.

Etymology. - Named for *Thylacinus cynocephalus* (Harris) (Thylacinidae), Australia's extinct marsupial carnivore.

Distribution. - Australia.

Biology. - Unknown.

Description (N=1). - BL 0.9 mm. BL/HTL = 4.1. Mesoscutal sculpturing longitudinally cellulate to striate with interstitial sculpturing primarily transverse. Forewing sparsely setose; AA absent; single setal track between CU1 and CU2; FWL/HTL = 2.8; FWL/FWW = 1.5; FWFS/FWW = 0.02; Max r-m to M/Min r-m to M = 1.6; MV/PM = 0.8; SV/MV = 1.2; MV length/MV width = 2.1. Hind wing width does not decrease immediately apical of hamuli; HWL/HWW = 5.8; HWFS/HWW = 0.4.

Male

Antenna: Club segments loosely joined; C/F = 1.7; F2/F1 = 0.9; APB absent on funicle; 1 PLS on each of F1-C2, 2 PLS on C3; 2-3 BPS on each of F1-C1, 1 BPS on C2 and C3; 13 FS on F1, 19 FS on F2, 17 FS on C1, 19 FS on C2, 15 FS on C3, 13 FS on C4; US absent on F1-C3.

Genitalia: Capsule relatively narrow, GL/GW = 4.0; GL/HTL = 1.3; ADA constricted abruptly in its posterior half, ADA/GL = 0.6; VP base likely hollow, VP/GL = 0.8; transverse hinge present at ca. 0.6 of GL; small paired appendages at apex of unknown homology; AI, AP, DR, PAR, VS absent.

Female

Unknown.

Other Material Examined. - None.

Comments. - The homology of the paired apical appendages found in *U. thylacinus* and *U. hercules* is unknown. Both are located ventrally, but those of *U. hercules* are smaller and somewhat more sclerotized. Other aspects of their male genitalia do not suggest a close relationship between these species. Nevertheless, SEM comparison of these structures should be given high priority when further material becomes available.

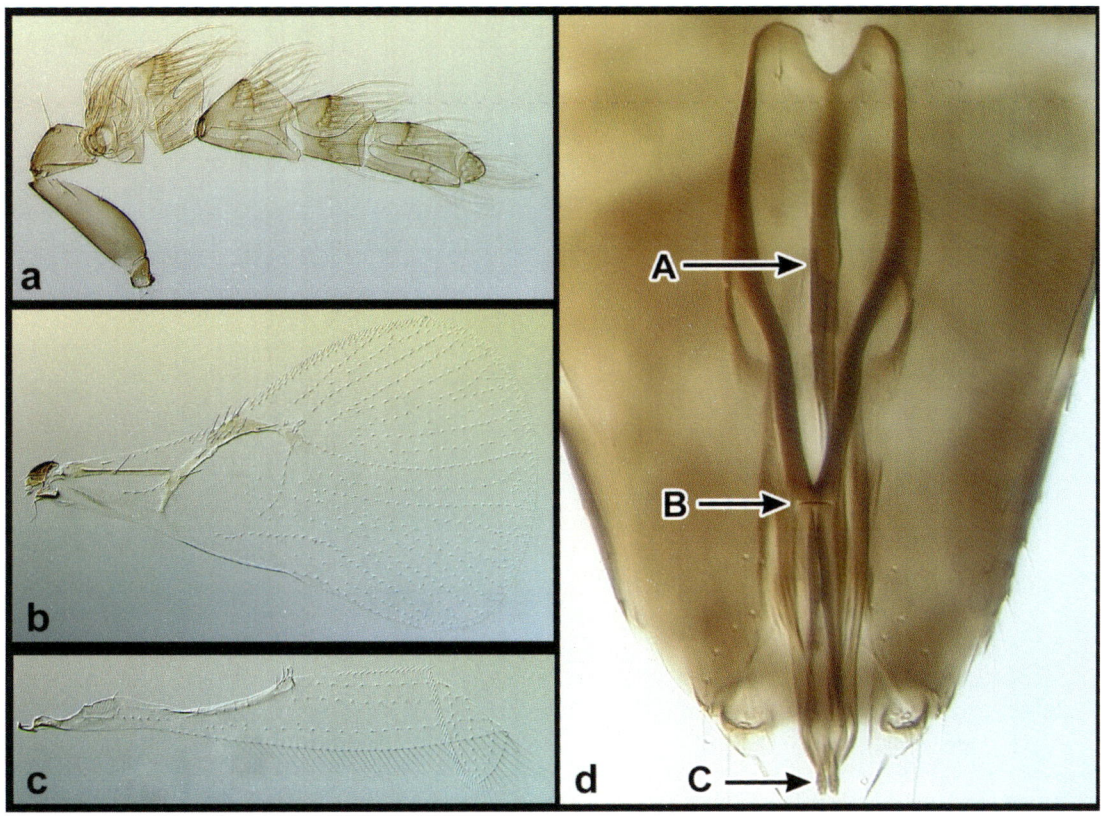

Figure 51. *Ufens thylacinus*, ♂. (a) antenna, lateral; (b) forewing, dorsal; (c) hind wing, dorsal; (d) genitalia, dorsal – arrows to {A} ventral process base, showing apparently hollow opening, {B} transverse hinge, {C} small paired appendages at apex of genitalia.

Ufens vectis Owen, new species
(Fig. 52)

Diagnosis. - Forewing densely setose with narrowly diverging setal tracks r-m to M and more than a single setal track between CU1 and CU2. Hind wing width decreasing immediately apical of hamuli. Mesoscutal sculpturing longitudinally cellulate to striate. Genitalia with apodemes present; parameres with terminal spine, widest at base, their base even with posterior edge of anterodorsal aperture; volsellae absent; ventral process short and rigid, its basal width < half of capsule width; dorsal projection curving precipitously posteroventrally; unidentifiable appendage closely appressed to venter of dorsal projection.

The genitalia of *U. vectis* are unique. Firstly, it is among the few species with aedeagal apodemes, but more importantly it is the only species in which the posterior portion of the genitalia (dorsal projection) dramatically projects posteroventrally. Of those species with apodemes, perhaps the most easily confused with *U. vectis* is *U. kurrajong*, from which *U. vectis* can also be separated by its shorter ventral process and parameres which are widest at the base rather than near the middle.

Types. - ◆Holotype ♂, Allotype ♀ (QM). **AUSTRALIA**: **Queensland**: Blackbutt, 9 km E, Blackbutt Creek, ix.22.1995, JDP, SW. Paratypes 3♂, 1♀, same data. (1♂ QM, remainder UCRC).

Etymology. - Latin for a lever, crow-bar, bar, or bolt - in reference to the large, curving dorsal projection of the male genitalia.

Distribution. - Australia, New Zealand, Indonesia, Papua New Guinea.

Biology. - Unknown.

Description. - BL 0.6 (0.5-0.8) mm. BL/HTL = 3.3 (3.0-3.6). Mesoscutal sculpturing longitudinally cellulate to striate with interstitial sculpturing longitudinal to transverse. Forewing densely setose; AA present; more than a single setal track between CU1 and CU2; FWL/HTL = 2.9 (2.8-3.0); FWL/FWW = 1.8 (1.8-1.9); FWFS/FWW = 0.07 (0.05-0.09); Max r-m to M/Min r-m to M = 2.0 (1.3-3.2); MV/PM = 1.1 (1.0-1.3); SV/MV = 0.9 (0.8-1.0); MV length/MV width = 3.4 (3.1-3.9). Hind wing width decreasing immediately apical of hamuli; HWL/HWW = 10.0 (9.2-11.2); HWFS/HWW = 1.2 (1.0-1.4).

Male

Antenna: Club segments loosely joined; C/F = 2.0 (1.8-2.1); F2/F1 = 1.2 (0.9-1.2); APB absent on funicle; 1 PLS on each of F1-C2, 2 PLS on C3; 2-4 BPS on each of F1-C1, 1-2 BPS on C2, 1 BPS on C3; 7-15 FS on F1, 10-17 FS on F2, 7-15 FS on C1, 9-17 FS on C2, 9-14 FS on C3, 7-12 FS on C4; US absent on F1-C3.

Genitalia: Capsule broad up to level of transverse hinge; dorsal projection long, thin, and curving posteroventrally; appendage of unknown homology closely appressed to venter of dorsal projection; GL/GW = 2.2 (2.0-2.4); GL/HTL = 0.7 (0.7-0.8); ADA/GL = 0.5 (0.4-0.6); PAR with terminal spine, widest at base, their base even with posterior edge of ADA; PAR/GL = 0.3 (0.2-0.4); VP short and rigid, its base < half of width of capsule, VP/GL= 0.2 (0.2-0.3); DR faint, extending ca. third of GL; AP/GL = 0.4 (0.3-0.4); transverse hinge immediately posterior of posterior edge of anterodorsal aperture; AI, VS absent.

Female

Antenna: C/F = 2.2 (2.1-2.3); F2/F1 = 1.0 (0.9-1.2); 1 APB on F1 and F2, 0-1 APB on C3; 1 PLS on F1, 1-2 PLS on F2 and C1, 2 PLS on C2, 3-4 PLS on C3; 5-6 BPS on F1, 3-4 BPS on F2, 2-4 BPS on C1, 1 BPS on C2 and C3; 0 FS on F1 and F2, 6-9 FS on C1, 9-11 FS on C2, 3-4 FS on C3; 1 UPP on C3; 5-13 US on F1, 7-15 US on F2, 6-13 US on C1, 0 US on C2, 2-5 US on C3.

Ovipositor: OL/HTL = 1.1 (1.0-1.3).

Other Material Examined

AUSTRALIA: **Australia Capital Territory**: Piccadilly Circus, 35°22'S, 148°48'E, 1240m. el., ii.1984, J. Lawrence, T. Wier, M-L. Johnson (1♂); Canberra, Black Mountain, ANIC, 35°16'S, 149°06'E, 11-19.i.1999, G. Gibson, MT (1♂).

Queensland: Weipa, 20.iv.1983, J. F. Donaldson, D-vac (1♂); Bluewater Ra., 50 km WNW Townsville, 700m el., 6-9.xii.1988, Monteith, Thompson and Hamlet, FIT (1♂); Townsville, nr. James Cook University, 15.iv.1988, E. C. Dahms and G. Sarnes (1♂); Charleville, Rd. to Augathella, 3.iii.1989, E. Dahms and G. Sarnes,

SW *Eucalyptus tesselaris, Aristida* spp., *Acacia excelsa* (1♂); Wangetti Beach Riple Range, 23 km SE Port Douglas, 31.iii.1991, JDP, SW (1♂); Mungana Rwy. Stn., NW Chillagoe, 17°06'25"S, 144°23'32"E, 8.iv.1992, E. C. Dahms and G. Sarnes (1♂); Tea Tree Cave, 4 km SE Chillagoe, 17°11'S, 144°34'E, 25.iv.1997, C. J. Burwell (1♂); Brisbane Forest Park, 27°25'04"S, 152°49'48"E, 30.xii.1997, N. Power, MT (1♂); Leichardt River Dam, 20°35'S, 139°35'E, 300m. el., 5-6.iii.2002, C. J. Burwell (1♂); Brisbane Forest Park, 27°25'04"S, 152°49'48"E, 16-23.x.1998, N. Power, MT (1♂); Lake Moondara, site 3, 20°34'S, 139°34'E, 340m. el., 6.iii.2002, C. J. Burwell (1♂); Great Sandy NP, off Rainbow Beach Rd. (43), 26°00.62'S, 153°02.80'E, 16.xii.2002, JBM and AKO, SW 1° grass/*Eucalyptus* forest (1♂, 1♀). **South Australia**: Brecon, 10 km S. Keith, 26.i.1982, A. D. Austin (1♂); 12 km E. Penong, 31°56'S, 133°08'E, 16.ix.1987, I. Naumann and J. Cardale, ex. ethanol (1♂). **New South Wales**: 100km S by E Broken Hill, 32°51'S, 141°37'E, 3-13.x.1988, E. D. Edwards, MT (1♂). **Northern Territory**: Timber Creek, 120 km W, 27.iii.1991, JDP, SW (1♂); Darwin, 53 km SSW, 12°52'10.5"S, 130°35'0.4"E, 24-28.xi.1997, M. Hoskins, MT in mango patch (2♂). **Western Australia**: Mining Camp, Mitchell Plateau, 14°49'S, 125°50'E, 9-19.v.1983, I. Naumann, J. Cardale, MT (1♂); Kununurra, 10 km N (Ivanhoe crossing), 24.iii.1991, JDP (1♂). **NEW ZEALAND**: **Aukland**: Birkenhead, xii.1980, J. F. Longworth, MT in second growth bush (1♂, 1♀); Birkenhead, i.1981, J. F. Longworth, MT in second growth growth (1♂, 2 ♀); Titirangi, ii.1981, P. A. Maddison, MT in garden (1♂). **"CO"** [likely Coromandel]: Watts Rock, 1200m. el., i.1981, J. S. Noyes and E. W. Valentine, SW tussock/grasses *Juncus*, *Sphagnum* (1♂). **Otago**: Dart Hut, 13-15.ii.1980, J. S. Dugdale (1♂, 1♀). **INDONESIA**: Jara Bogor, S. G. Compton, MT (1♂) [No date specified, but donated to SAM 1994]; Krakatoa, Anak, 13.ix.1984, S. G. Compton, SW (1♂). **PAPUA NEW GUINEA**: **Central Province**: SDA College, 25 km NE Port Moresby, 31.xii.1985, G. Gordh, SW (1♂, 1♀); 15 km SE Port Moresby; 1.i.1986, G. Gordh, SW *Eucalyptus* grassland (1♂, 1♀). **East New Britain Province**: Bainings Mountains, Raunsepna, 14-21.iv.1999, L. Leblanc and C. Mitparingi, MT (1♀); Bainings Mountains, DPI base camp, 04°26'36"S, 151°49'02"E, 15.ix-14.x.1999, A. Mararuai and M. Kalamen (1♀). **Madang Province**: Awar Airfield, st. 1350, 18.vi.1982, P. Grootaert (1♂).

Comments. - Other than several specimens of *U. foersteri*, *U. vectis* is the only *Ufens* known from both within and outside of Australia.

A male of this species can be found on a slide located at the USNM containing "*Ufens flavipes* Girault ♀" under a complete coverslip. The male *U. vectis* is located adjacently under a half coverslip, and is designated "*Ufens* sp."

Molecular data for *U. vectis* were presented in Owen et al. (2007) as *Ufens* sp. 9, and can be found under Genbank accession numbers AY623538 (28S-D2+D3) and AY940379 (18S).

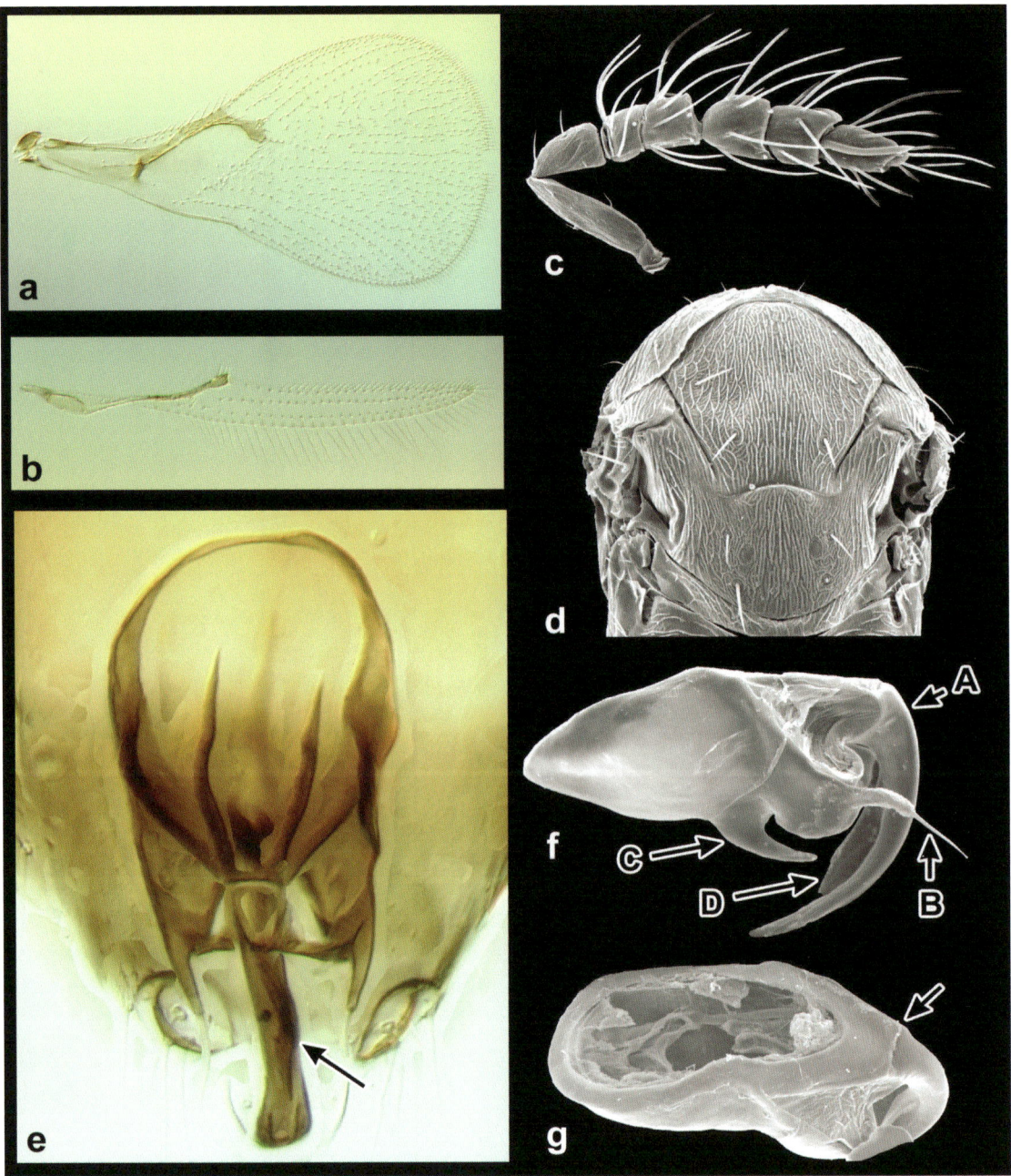

Figure 52. *Ufens vectis*, ♂. (a) forewing, dorsal; (b) hind wing, dorsal; (c) antenna, medial; (d) mesosoma, dorsal; (e) genitalia, dorsal – arrow to unidentified appendage closely associated with dorsal projection (visible as darkened sinuous area); (f) genitalia, lateral – arrows to {A} dorsal projection, {B} paramere, {C} ventral projection, {D} appendage closely appressed to dorsal projection; (g) genitalia, dorsal – arrow to transverse hinge.

Nomina dubia

The following names either are based on females only or associated males have not been located. Because in all cases these females cannot be distinguished from those of known species these nominal taxa are unidentifiable and treated here as *nomina dubia* (cf. Table 2).

Ufens albitibiae Girault, 1915

U. albitibiae Girault, 1915: p. 145.

Dahms, 1983: p. 35 (type material described).

Types. - Holotype ♀. **AUSTRALIA**: **Queensland**: Mackay (QM).

Comments. - According to Girault (1915) this species was described from "one female captured by sweeping miscellaneous vegetation along the banks of the Pioneer River. October 15, 1911...Mackay, Queensland." The specimen presumed to be the holotype of this species was not examined, but as listed in Dahms (1983) (as slide 1), it is confirmed to be a female with the head separated and one antenna missing, though the body is intact (C. Burwell, pers. comm.). The specimen listed in Dahms (1983) as slide 2 has been examined and is a laterally mounted female in relatively good condition.

Ufens alami Yousuf and Shafee, 1987

U. alami Yousuf and Shafee, 1987: p. 74.

Types. - ◆Holotype ♀ (BMNH). **INDIA**: **Uttar Pradesh**: Aligarh, 30.i.1985, M. Yousuf, ex. eggs of *Oxyrachis tarandus* (Fabricius) (Membracidae).

Comments. - *U. alami* was apparently decribed from a single female. The specimen is on two slides, both labelled as holotype ♀; one with the body and the other with a forewing and antenna. The dissected forewing is ripped and the club is separated from the rest of the antenna. The rest of the body on the second slide is difficult to discern. This species was recognized as differing "from all known species of *Ufens* Gir

ault by having antenna with first funicle segment more than one half the length of the second; fore wings with marginal vein one-half the length of the stigmal vein."

In fact, it is common for *Ufens* females to have the first funicle segment greater than one-half the length of the second. The forewing in general seems unremarkable. It is sparsely setose, but this is also common of many *Ufens* species. Males are unknown, rendering current recognition of this species impossible. However, as the host of this specimen is known, there is the possibility that males could be reared and the identity of this taxon verified.

Ufens angustipennis Yousuf and Shafee, 1987

<u>*U. angustipennis* Yousuf and Shafee, 1987: pp. 77-78.</u>
Types. - ◆Holotype ♀ (BMNH). **INDIA: Uttar Pradesh**: Aligarh, 1.ii.1985, M. Yousuf, ex. eggs of *Oxyrachis* sp. (Membracidae). Paratypes 1♀, 1♂ same data.
Comments. - This species was reported to be closely related to *U. foersteri* Kryger, but differing by a forewing with the stigmal vein longer than the marginal, RS1 track reaching beneath premarginal vein, and the costal cell broad with two rows of setae. No males of this species were found in the BMNH, though one is listed in the description. The location of this male paratype is unknown. Unfortunately, this male was not described by Yousuf and Shafee. Contrary to statements by these authors, the forewings of *U. angustipennis* are unremarkable for *Ufens*. Because the male paratype has not been located and is not described, identification of this species is impossible. However, even if the male paratype is not located, there is the possibility that males could be reared and the identity of this taxon verified.

Ufens binotatus Girault 1915

<u>*U. binotatus* Girault, 1915: pp. 145-146.</u>
<u>Dahms, 1983: p. 141 (type material described).</u>
Types. - ◆Type ♀ (QM). **AUSTRALIA: Queensland**: "Ufens binotatus Girault, ♀ type, 3439, Lathromeromella longiciliata Gir., ♂", "Paratype, T. 3490, E. C. D. 1984".
Comments. - This species was described from "one female caught on native grass in forest. April 4, 1914...Gordonvale (Cairns), Queensland." Girault (1915) also reported a second female which was collected on May 13 in the same locality. However, this second female seems to have been lost and was not found by Dahms (1983). The head of the type examined is separated from the body, though the antenna and forewing can be seen reasonably well. The forewing is moderately setose, with an ovipositor slightly longer than the hind tibial length. Overall, the specimen is unremarkable and unidentifiable.

Ufens breviclavata Yousuf and Shafee, 1991

<u>*U. breviclavata* Yousuf and Shafee, 1991: p. 59.</u>
Types. - ◆Holotype ♀ (BMNH). **INDIA: Uttar Pradesh:** Moradabad, 18.ix.1985, M. Yousuf, ex. *Oxyrachis tarandus* (Membracidae).
Comments. - *U. breviclavata* was apparently described from a single female. No locality was mentioned in the original description (Yousuf and Shafee 1991). It was hypothesized to be related to *Mirufens albiscutellum* Khan and Shafee (as *Ufens albiscutellum*), but differing "in having fore wings with marginal vein very long, division of funicular segments inconspicuous." This specimen is clearly an *Ufens*

and obviously unrelated to *M. albiscutellum*. The forewings of this specimen are unremarkable for *Ufens*, including the marginal vein, which seems typical. However, as the host of this specimen is known, there is the possibility that males could be reared and the identity of this taxon verified.

Ufens gurgaonensis Yousuf and Shafee, 1987

U. gurgaonensis Yousuf and Shafee, 1987: pp. 75-77.

Types. - ◆Holotype ♀ (BMNH). **INDIA: Haryana:** Gurgaon, 1.x.1984, A. K. Chisti, SW.

Comments. - *U. gurgaonensis* was described from a single female. The specimen is on two slides, both labelled as holotype ♀; one with the body and the other with a forewing and antenna. It was hypothesized to be closely related to *Mirufens brevifuniculata* Khan and Shafee (as *Ufens brevifuniculata*), but was distinguished by the more densely setose forewings and the cylindrical funicle. As a true *Ufens*, it is not related to *M. brevifuniculata*. Neither its forewing setation nor shape of the funicle are remarkable.

Ufens jaipurensis Yousuf and Shafee, 1987

U. jaipurensis Yousuf and Shafee, 1987: pp. 80-82.

Types. - ◆Holotype ♀ (BMNH). **INDIA: Rajasthan:** Jaipur, 16.x.1985, M. Yousuf, SW.

Comments. - *U. jaipurensis* was described from a single female. The specimen is on two slides, both labelled as holotype ♀; one with the body and the other with a forewing and antenna (somewhat damaged). It was hypothesized to be closely related to *Mirufens magniclavata* Khan and Shafee (as *Ufens magniclavata*), but differing by having the first funicle segment less than half the length of the second, a club 3.5 x as long as wide, and a basal vein track of 4 setae. As a true *Ufens*, it is not related to *M. magniclavata*. The antenna is unremarkable.

Ufens latipennis Yousuf and Shafee, 1987

U. latipennis Yousuf and Shafee, 1987: pp. 78-80.

Types. - ◆Holotype ♀ (BMNH). **INDIA: Uttar Pradesh:** Aligarh, 6.viii.1985, M. Yousuf, ex. eggs of membracids.

Comments. - *Ufens latipennis* was described from a single female. The holotype is on two slides, both labelled as holotype ♀; one with the body and the other with a forewing and antenna. It was considered closely related to *U. africana* Viggiani, but was distinguishable by having broader forewings and an antenna with the scape less than 3x as long as wide. These traits are not distinctive as both are within the range of variation found in *U. foersteri*, the senior synonym of *U. africana*, and most other species for that matter.

Ufens luna Girault, 1911b

U. luna Girault, 1911b: pp. 198-199.
Girault, 1916 (redescription): pp. 205-206.
Dahms, 1984: p. 777 (type material described).
Types. - ◆Holotype ♀ (USNM). **AUSTRALIA: Western Australia**: "923. Perth W. Austr., G. Compere, Ufens luna Girault ♀, Type 13794."
Comments. - According to Girault (1911b) this species was "described from a single female specimen received from Dr. L. O. Howard, It is mounted in balsam, and labeled: '923. Perth. W. Austr. G. Compere.'" The 1916 redescription appears to have been from the same specimen. The head is separated from the body, though both antennae are complete and attached to the head. The only forewing is ripped into two pieces, and the only hind wing is well-removed from the body. Overall, the specimen is unremarkable.

Ufens piceipes Girault, 1912

U. piceipes Girault, 1912: pp. 71-72.
Dahms, 1986: p. 409 (type material described).
Types. - ◆Type ♀ (QM). **AUSTRALIA: Queensland**: "Queensland Museum. 3381. 3437. TYPE, Hy/777, Hy/797", "Ufens piciepes Girault, 777, *Aphelinoidea howardii* Gir. From windows of a barn, Roma, Qld., 6 Oct., 1911, AAG, Types 3437. 3381."
Comments. - According to Girault (1912) this species was "described from two females captured from the pane of a window in a barn. State Farm, Roma, Queensland, 6 October, 1911." These two females were designated as syntypes, and placed on separate slides. Slide 1 [as designated by Dahms (1986)] has been examined and consists of a female with the head detached, the body intact and dorsoventrally flattened, and all wings folded over. Slide 2 was reported to contain the second syntype. This slide lists six original specimens of various genera, and only five are currently present. Four of these five are readily associated with their appropriate genus. However, the fifth is located in an area of the slide where the coverslip has been removed or fallen off, rendering it unidentifiable. Based upon the original constituents of the slide, it is either *Ufens* or *Aphelinoidea*.

Ufens pretiosus (Girault, 1913), *new combination*

Ufensia pretiosa Girault, 1913: p. 102.
Ufensia pretiosa Girault, 1914: p. 118 (redescription).

Dahms, 1986: p. 425 (type material described).

Types. - ◆Holotype ♀ (QM). **AUSTRALIA: Queensland**: Slide 1: "*Oligosita sacra* Girault ♀, window, Nelson, N. Q., 10 Oct., 1912.", [on reverse] "Queensland Museum TYPE Hy/1173 ♀."
Slide 2: "3763", "3762", "Type ♀, *Perissopterus: argenticorpus* Gir., *angeloni* Girault, *Ufensia pretiosa*, Qsld., P. T. O." (QM).
Comments. - [Described as *Ufensia pretiosa*, in both 1913 and 1914.]
Girault (1914) indicated that this species was "described from a single female captured by sweeping grass in a forest near Nelson, N. Q., October 10, 1912." Girault (1913, 1914) mentioned a single specimen in his descriptions identified with "Hy/1173". Therefore, the specimen on the second slide does not have type status, and is likely subsequently identified material. The holotype is somewhat easily seen and is mounted laterally. It has a sparsely setose forewing and a long ovipositor, nearly three times the hind tibial length. The specimen on the second slide is difficult to see (as the coverslip is cracked), and is mounted laterally. The ovipositor of the second specimen is not as long, though it is difficult to measure. The length of the ovipositor of the holotype is certainly distinctive, though specimens with even longer ovipositors are known (AKO, unpublished; cf. Fig. 9). Girault (1913) considered the length and exsertion of the ovipositor to be of such significance that he erected the new genus *Ufensia* to accommodate it.

Ufensia was later synonimized by Doutt and Viggiani (1968), only to be subsequently resurrected by Viggiani (1972), to include *U. africana* (herein considered a synonym of *U. foersteri*), based upon the distinctness of its male genitalia. *Ufensia* is herein synonomized under *Ufens* based on both molecular (Owen et al. 2007) and morphological (cf. *Ufens* generic description above) information.

In spite of its long ovipositor, the relationship of the *U. pretiosa* holotype to other *Ufens* species cannot be resolved. Firstly, there is the problem that so many of the species known from Queensland do not have unambiguously associated females. Secondly, its antennae are badly shriveled, making it very difficult to determine setal counts, etc. Finally, the forewing is readily visible and somewhat sparsely setose, but unremarkable for *Ufens*. There remains the possibility that this specimen could be a female of *U. foersteri*, which is also known in small numbers from Australia. This possibility is intriguing as most other previously recognized *Ufensia* species (*U. dilativena* is the exception) are synonymyzed under *U. foersteri*. Some specimens of *U. foersteri* are known with similar ovipositor lengths and the forewing setation is very compatible. However, in lieu of further information, a positive association cannot be made.

Ufens quadrifasciatus Girault, 1915

<u>*U. quadrifasciatus*</u> Girault, 1915: p. 145.
Dahms, 1986: p. 455 (type material described).
Types. - ◆Holotype ♀. **AUSTRALIA**: **Queensland**: *"Ecthrobacomyia niveipes* ♀,
Ufens quadrifasciatus Girault ♀, type."* (QM)
Comments. - Girault (1915) indicated that this species was "described from one
female captured in jungle pocket, April 2, 1914…Gordonvale (Cairns),
Queensland." The specimen is mounted laterally, with a single forewing discernible.
It has an ovipositor that is slightly longer than the hind tibia and sparse forewing
setation. Its features do not allow identification.

Ufens singularis Yousuf and Shafee, 1987

<u>*U. singularis*</u> Yousuf and Shafee, 1987: pp. 74-75.
Types. - ◆Holotype ♀ (BMNH). **INDIA**: **Uttar Pradesh:** Aligarh, 14.v.1985, M.
Yousuf, ex. eggs of *Oxyrachis* sp. (Membracidae).
Comments. - *U. singularis* was described from a single female. The specimen is on
two slides, both labelled as holotype ♀; one with the body and the other with a
forewing and antenna. It was hypothesized to be closely related to *U. dilativena*
Nowicki and recognized by antennae with the first funicular segment one quarter the
length of the second, and the RS1 extending beneath the radial process. It cannot be
associated with known species. With respect to the purported distinguishing
features, considerable intraspecific variation in the relative length of funicle
segments is known in other species, and the RS1 length reported is characteristic of
many other species.

References

Al-Wahaibi, A. K. (2004). Studies on two Homalodisca species (Hemiptera: Cicadellidae) in southern California: biology of the egg stage, host plant and temporal effects on oviposition and associated parasitism, and the biology and ecology of two of their egg parasitoids, Ufens A and Ufens B (Hymenoptera: Trichogrammatidae). Ph.D. Thesis. Department of Entomology, University of California, Riverside, California.

Al-Wahaibi, A. K., A. K. Owen and J. G. Morse (2005). Description and behavioural biology of two *Ufens* species (Hymenoptera: Trichogrammatidae), egg parasitoids of *Homalodisca* species (Hemiptera: Cicadellidae) in southern California. *Bulletin of Entomological Research* **95**: 275-288.

Amornsak, W., B. Cribb and G. Gordh (1998). External morphology of antennal sensilla of *Trichogramma australicum* Girault (Hymenoptera: Trichogrammatidae). *International Journal of Insect Morphology and Embryology* **27**: 67-82.

Ashmead, W. H. (1888). Descriptions of some new North American Chalcididae. *The Canadian Entomologist* **20**: 101-107.

Babcock, C. S., J. M. Heraty, P. J. De Barro, F. Driver, S. Schmidt (2001). Preliminary phylogeny of Encarsia Förster (Hymenoptera: Aphelinidae) based on morphology and 28S rDNA. *Molecular Phylogenetics and Evolution* **18**: 306-323.

Blood, B. N. (1923). Notes on Trichogrammatinae taken around Bristol. *Proceedings of the Bristol Naturalists' Society (Annual Report)* **5**: 253-258.

Blood, B. N. and J. P. Kryger (1928). New genera and species of Trichogrammidae with remarks upon the genus *Asynacta* [Hym. Trichogr.]. *Entomologiske Meddelelser* **16**: 203-222.

Blua, M. J., P. A. Phillips and R. A. Redak (1999). A new sharpshooter threatens both crops and ornamentals. *California Agriculture* **53**: 22-25.

Dahms, E. C. (1983). A checklist of the types of Australian Hymenoptera described by Alexandre Aresene Girault: II. Preamble and Chalcidoidea species A-E with advisory notes. *Memoirs of the Queensland Museum* **21**: 1-255.

Dahms, E. C. (1984). A checklist of the types of Australian Hymenoptera described by Alexandre Arsene Girault: III. Chalcidoidea species F-M with advisory notes. *Memoirs of the Queensland Museum* **21**: 579-842.

Dahms, E. C. (1986). A checklist of the types of Australian Hymenoptera described by Alexandre Arsene Girault: IV. Chalcidoidea species N-Z and genera with advisory notes plus addenda and corrigenda. *Memoirs of the Queensland Museum* **22**: 319-739.

Doutt, R. L. (1973). The genus *Paratrichogramma* Girault (Hymenoptera: Trichogrammatidae). *The Pan-Pacific Entomologist* **49**: 192-196.

Doutt, R. L. and G. Viggiani (1968). The classification of the Trichogrammatidae (Hymenoptera: Chalcidoidea). *Proceedings of the California Academy of Sciences* (4th ser.) **35**: 477-586.

Farris, J. S. (1969). A successive approximations approach to character weighting. *Systematic Zoology* **18**: 374-385.

Fursov, V. N. (1994). A world review of *Uscana* species (Hymenoptera, Trichogrammatidae), potential biological control agents of bruchid beetles (Coleoptera, Bruchidae). *Bulletin of the Irish Biogeographical Society* **18**: 2-12.

Gibson, G.A.P. (1997). Chapter 2. Morphology and Terminology, pp. 16-44. *In*: Gibson, G. A. P., J. T. Huber and J. B. Woolley (eds.). *Annotated Keys to the genera of Nearctic Chalcidoidea (Hymenoptera)*. National Research Council Research Press, Ottawa, Canada: 794 pp.

Girault, A. A. (1911a). Descriptions of nine new genera of the chalcidoid family Trichogrammatidae. *Transactions of the American Entomological Society* **37**: 1-83.

Girault, A. A. (1911b). Two new species of Trichogrammatidae from the United States and West Australia. *The Entomologist* **44**: 198-199.

Girault, A. A. (1912). Australian Hymenoptera Chalcidoidea - I. *Memoirs of the Queensland Museum* **1**: 66-116.

Girault, A. A. (1913). Australian Hymenoptera Chalcidoidea - I. Supplement. *Memoirs of the Queensland Museum* **2**: 101-106.

Girault, A. A. (1914). Descriptions of new chalcid-flies. *Proceedings of the Entomological Society of Washington* **16**: 109-119.

Girault, A. A. (1915). Australian Hymenoptera Chalcidoidea - I. Second Supplement. *Memoirs of the Queensland Museum* **3**: 142-153.

Girault, A. A. (1916). Australian Hymenoptera Chalcidoidea. General supplement. *Memoirs of the Queensland Museum* **5**: 205-230.

Girault, A. A. (1939). Five new generic names in the Chalcidoidea (Australia). *Ohio Journal of Science* **39**: 324-326.

Gordh, G. and J. Hall (1979). Critical point drying: application of the physics of the PVT surface to electron miscroscopy. *American Journal of Physics* **43**: 414-419.

Harris, R. A. (1979). A glossary of surface sculpturing. *California Department of Agriculture, Bureau of Entomology, Occasional Papers,* Sacramento California **28**: 1-31.

Heraty, J. M., J. B. Woolley and D. C. Darling (1997). Phylogenetic implications of the mesofurca and mesopostnotum in Chalcidoidea (Hymenoptera), with emphasis on Aphelinidae. *Systematic Entomology* **3**: 241-277.

Heraty, J. and D. Hawks (1998). Hexamethyldisilazane - a chemical alternative for drying insects. *Entomological News* **109**: 369-374.

Kryger, J. P. (1918). The European Trichogramminae. *Entomologiske Meddelelser* **12**: 257-354.

Kryger, J. P. (1932). One new genus and species, and three new species of Trichogramminae from Egypt with remarks upon *Neocentrobia hirticornis, Alaptus minimus,* and *Trichogramma evanescens. Bulletin de la Société Royale Entomologique d'Egypte* **16**: 38-44.

Li, L.-Y. (1994). Worldwide use of *Trichogramma* for biological control of different crops: A survey, pp. 37-53. *In* E. Wajnberg and S. Hassan (eds.) Biological Control with Egg Parasitoids, CAB International, UK.

Lin, N. (1993). On Chinese species of the genus *Ufens* Girault, with descriptions of new species and a new record (Hymenoptera: Trichogrammatidae). *Wuyi Science Journal* **10**: 51-59.

Lin, N. (1994). Systematic Studies of Chinese Trichogrammatidae. Contributions of the Biological Control Research Institute. Fujian Agricultural and Forestry University, Fuzhou, China, Special Publication No. 4, 362 pp.

Lin, N. (2002). A new species of *Hispidophila* and description of the female of *Ufens rimatus* (Hymenoptera: Trichogrammatidae), parasitoids of *Sophonia* leafhoppers (Homoptera: Cicadellidae). *Acta Zootaxonomica Sinica* **27**: 347-350.

Nagaraja, H. (1978). Studies on the *Trichogrammatoidea* (Hymenoptera: Trichogrammatidae). *Oriental Insects* **12**: 489-530.

Neto, L. and B. Pintureau (1997). Review of the genus *Mirufens* Girault (Hymenoptera: Trichogrammatidae). *Entomological Problems* **28**: 141-148.

Nikol'skaya, M. N. (1952). *Family Trichogrammatidae*. The Chalcid Fauna of U. S. S. R. M. N. Nikol'skaya, ed. Zoological Institute of the Academy of Sciences of the U.S.S.R. **44**: 524-555. [Published for the National Science Foundation, Washington, D. C., 1963. Translated from Russian by the Israel Program for Scientific Translation, Jerusalem].

Nikol'skaya, M. N. and V. A. Trjapitzin (1987). *Family Trichogrammatidae (Trichogrammatids)*. Keys to the Insects of the European Part of USSR. G. S. Medvedev, ed. New Delhi, Amerind Publishing Co. III, part II: 501-513.

Nowicki, S. (1935). Descriptions of new genera and species of the family Trichogrammatidae (Hym. Chalcidoidea) from the palearctic region, with notes - I. *Zeitschrift fur Angewandte Entomologie* **21**: 566-596.

Nowicki, S. (1940). Descriptions of new genera and species of the family Trichogrammatidae (Hym. Chalcidoidea) from the Palearctic region, with notes - Supplement. *Zeitschrift für Angewandte Entomologie* **26**: 624-663.

Noyes, J. S. (1985). Chalcidoids and biological control. *Chalcid Forum* (newsletter) **5**: 5-13.

Noyes, J. S. (2002). Catalogue of the Worl Chalcidoidea (2001 – 2nd edition). Compact disc by Taxapad, Vancouver, Canada and the Natural History Museum, London.

Olson, D. M. and D. A. Andow (1993). Antennal sensilla of female *Trichogramma nubilale* (Ertle and Davis) (Hymenoptera: Trichogrammatidae) and comparisons with other parasitic Hymenopotera. *International Journal of Insect Morphology and Embryology* **22**: 507-520.

Peck, O. (1963). *Trichogrammatidae*. A Catalogue of the Nearctic Chalcidoidea (Insecta: Hymenoptera). O. Peck, ed. The Canadian Entomologist. Supplement **13**: 51-86.

Olson, D. M., E. Dinerstein, E. D. Wikramanaya, N. D. Burgess, G. V. N. Powell, E. C. Underwood, J. A. D'Amico, I. Itoua, H. E. Strand, J. C. Morrison, C. J. Loucks, T. F. Allnutt, T. H. Ricketts, Y. Kura, J. F. Lamoruex, W. W. Wettengel, P. Hedao, and K. R. Kassem (2001). Terrestrial ecoregions of the world: A new map of life on earch. *Bioscience* **51**: 933-938.

Owen, A. K., J. George, J. D. Pinto and J. M. Heraty (2007). A molecular phylogeny of the Trichogrammatidae (Hymenoptera: Chalcidoidea), with an evaluation of the utility of their male genitalia for higher level classification. *Systematic Entomology* **32**: 227-251.

Pinto, J. D. (1990). The genus *Xiphogramma*, its occurrence in North America, and remarks on closely related genera (Hymenoptera: Trichogrammatidae). *Proceedings of the Entomological Society of Washington* **92**: 538-543.

Pinto, J. D. (1997). Trichogrammatidae, pp. 726-752. *In* G. A. P. Gibson, J. T. Huber and J. B. Woolley (eds.) Annotated Keys to the Genera of Nearctic Chalcidoidea.. Ottawa, NRC Research Press: 794 pp.

Pinto, J. D. (1999). Systematics of the North American species of *Trichogramma* Westwood (Hymenoptera: Trichogrammatidae). Washington D. C., The Entomological Society of Washington **22**: 287 pp.

Pinto, J. D. (2006). A review of the New World Genera of Trichogrammatidae (Hymenoptera). *Journal of Hymenoptera Research* **15**: 38-163.

Pinto, J. D. and J. George (2004). *Kyuwia*, a new genus of Trichogrammatidae (Hymenoptera) from Africa. *Proceedings of the Entomological Society of Washington* **106**: 531-539.

Pinto, J. D., E. R. Oatman and G. R. Platner (1986). *Trichogramma pretiosum* and a new cryptic species occurring sympatrically in southwestern North America (Hymenoptera: Trichogrammatidae). *Annals of the Entomological Society of America* **79**: 1019-1028.

Pinto, J. D. and A. K. Owen (2004). *Adryas*, a new genus of Trichogrammatidae (Hymenoptera: Chalcidoidea) from the New World tropics. *Proceedings of the Entomological Society of Washington* **106**: 905-922.

Pinto, J. D., G. R. Platner and R. Stouthamer (2003). The systematics of the *Trichogramma minutum* species complex (Hymenoptera: Trichogrammatidae), a group of important North American biological control agents: the evidence from reproductive compatibility and allozymes. *Biological Control* **27**: 167-180.

Pinto, J. D. and R. Stouthamer (1994). Systematics of the Trichogrammatidae with emphasis on *Trichogramma*. Biological Control with Egg parasitoids. E. Wajnberg and S. A. Hassan. Wallingford, CAB International: 1-36.

Pinto, J. D., R. Stouthamer and G. R. Platner (1997). A new cryptic species of *Trichogramma* (Hymenoptera: Trichogrammatidae) from the Mojave Desert of

California as determined by morphological, reproductive and molecular data. *Proceedings of the Entomological Society of Washington* **99**: 238-247.

Pinto, J. D., R. K. Velten, G. R. Platner and E. R. Oatman (1989). Phenotypic plasticity and taxonomic characters in *Trichogramma* (Hymenoptera: Trichogrammatidae. *Annals of the Entomological Society of America* **82**: 414-425.

Pinto, J. D. and G. Viggiani (1991). A taxonomic study of the genus *Ceratogramma* (Hymenoptera: Trichogrammatidae). *Proceedings of the Entomological Society of Washington* **93**: 719-732

Pinto, J. D. and G. Viggiani (2004). A review of the genera of Oligositini (Hymenoptera: Trichogrammatidae) with a preliminary hypothesis of phylogenetic relationships. *Journal of Hymenoptera Research* **13**: 269-294.

Platner, G. R., R. K. Velten, M. Planoutene, and J. D. Pinto (1999). Slide-mounting techniques for Trichogramma (Trichogrammatidae) and other minute parasitic Hymenoptera. *Entomological News* **110**: 56-64.

Powers, N. R. (1973). The biology and host plant relations of *Homalodisca lacerta* (Fowler) in southern California. MS thesis, Department of Biology, California State University. San Diego, California.

Scotland, R. W., R. G. Olmstead and J. R. Bennett (2003). Phylogeny reconstruction: the role of morphology. *Systematic Biology* **52**: 539-548.

Smith, S. M. (1996). Biological control with *Trichogramma*: advances, successes, and potential of their use. *Annual Review of Entomology* **41**: 375-406.

Soika, W. (1931). *Stephanotheisa* Eine Neue Trichogramminengattung. *Natuurhistorisch Maanblad* **20**: 111-112.

Swofford, D. L. (2001). PAUP*: Phylogenetic Analysis Using Parsimony (*and Other Methods). Sunderland, MA, Sinauer Associates.

Timberlake, P. H. (1927). New species of Hawaiian chalcid-flies (Hymenoptera) - II. *Proceedings of the Hawaiian Entomological Society* **6**: 517-528.

Triapitsyn, S. V. (2003). Taxonomic notes on the genera and species of Trichogrammatidae (Hymenoptera) - egg parasitoids of proconiine sharpshooters (Hemiptera: Clypeorryncha: Cicadellidae: Proconiini) in southeastern USA. *Transactions of the American Entomological Society* **129**: 245-265.

Triapitsyn, S. V., L. G. Bezark and D. J. W. Morgan (2002). Redescription of *Gonatocerus atriclavus* Girault (Hymenoptera: Mymaridae), with notes on other egg parasitoids of sharpshooters (Homoptera: Cicadellidae: Prociini) in Northeastern Mexico. *Pan-Pacific Entomologist* **78**: 34-42.

Triapitsyn, S. V. and M. S. Hoddle (2001). Search for and collect egg parasitoids of the glassy-winged sharpshooter in southeastern USA and northeastern Mexico. Proceedings of the Pierce's disease research symposium, 5-7 December 2001, Coronado Island Marriot Resort, San Diego, California, California Department of Food and Agriculture, Sacramento, California.

Triapitsyn, S. V. and M. S. Hoddle (2002). Search for and collect egg parasitoids of the glassy-winged sharpshooter in southeastern USA and northeastern Mexico. Proceedings of the Pierce's disease research symposium, 15-18 December 2002, Coronado Island Marriot Resort, San Diego, California, California Department of Food and Agriculture, Sacramento, California.

Triapitsyn, S. V., R. F. Mizell, J. L. Bossart and C. E. Carlton (1998). Egg parasitoids of *Homalodisca coagulata* (Homoptera: Cicadellidae). *Florida Entomologist* **81**: 241-243.

Varela, L. G., R. J. Smith and P. A. Phillips (2001). Pierce's disease. University of California, Oakland, California, University of California, Division of Agriculture & Natural Resources, Publication No. 21600.

Velten, R. K. and J. D. Pinto (1990). *Soikiella* Nowicki (Hymenoptera: Trichogrammatidae): Occurrence in North America, description of a new species, and association of the male. *Pan-Pacific Entomologist* **66**: 246-250.

Viggiani, G. (1971). Ricerche sugli Hymenoptera Chalcidoidea XXVIII. Studio morfologico comparativo dell'armatura genitale esterne maschile dei Trichogrammatidae. *Bollettino del Laboratorio di Entomologia Agraria 'Filippo Silvestri'* **29**: 181-222.

Viggiani, G. (1972). Ricerche sugli Hymenoptera Chalcidoidea XXXVI. Nuovi Tricogrammatidi africani. *Bollettino del Laboratorio di Entomologia Agraria 'Filippo Silvestri'* **30**: 158-164.

Viggiani, G. (1988). *Ufensia minuta* sp. n., (Hymenoptera: Trichogrammatidae), ooparassitoide di *Reuteria marqueti* Puton (Hemiptera: Miridae), con note sulle specie paleartiche del genere *Ufensia* Girault. *Bollettino del Laboratorio di Entomologia Agraria 'Filippo Silvestri'* **45**: 15-21.

Vincent, D. L. and C. Goodpasture (1986). Three new species of *Trichogramma* (Hymenoptera: Tichogrammatidae) from North America. *Proceedings of the Entomological Society of Washington* **88**: 491-501.

Wiens, J. J. (2004). The role of morphological data in phylogeny reconstruction. *Systematic Biology* **53**: 653-661.

Yousuf, M. and S. A. Shafee (1987). Taxonomy of Indian Trichogrammatidae (Hymenoptera: Chalcidoidea). *Indian Journal of Systematic Entomology* **4**: 55-200.

Yousuf, M. and S. A. Shafee (1991). Two new species of Trichogrammatidae (Hymenoptera: Chalcidoidea) from India. Proceedings of the 78th Indian Scientific Congress Part III (Advance Abstracts).

CPSIA information can be obtained
at www.ICGtesting.com
Printed in the USA
258131LV00006B